Your Guide to NATURE RESERVES in LINCOLNSHIRE

S.E. Crooks

FOREWORD BY
Professor David Bellamy

FOURTH EDITION

THE wildlife TRUSTS

The Lincolnshire Trust
for nature conservation

Headquarters:
Banovallum House
Manor House Street
Horncastle
Lincolnshire LN9 5HF
Tel: (01507) 526667
Fax: (01507) 525732
e-mail: lincstrust @
cix.compulink.co.uk
World-wide web site:
http: //www.enterprise.net/
wildlife/lincstrust

First published 1974
Second edition 1981
Third edition 1989
Fourth edition 1997

Printed and bound by
GW Belton Ltd, Heaton Street,
Gainsborough, Lincolnshire
DN21 2ED
Tel: (01427) 612291
Fax: (01427) 810520

The Lincolnshire Trust for Nature Conservation is a company limited by guarantee (Reg. in England no. 461863) and is a registered charity (Reg. charity no. 218895)

A catalogue record for the title is available from the British Library

ISBN 0 9501420 8 5

Contents

Foreword

Warm summer meadows buzzing with bees and full of wildflowers, freshwater streams teeming with life, ancient bluebell woods echoing with spring bird song – memories are made from such wonders! Where can you go to explore the wild world like we did when we were children, and where can the children of today go?

Looking after special places where nature can be kept safe, 'in reserve', is just part of the valuable work carried out by the Wildlife Trusts nationally, with over 2000 nature reserves now flourishing in their care. The Lincolnshire Trust for Nature Conservation was in there protecting wild places long before 'green' issues became fashionable. This book gives full information on the Trust's 104 nature reserves, where they are, how to find them and their importance in enhancing the Great Biodiversity of Life.

Nature reserves were never intended to be the *only* places where wildlife survives, but such has been the extent of habitat loss in the wider countryside that, in many cases, Lincolnshire Trust nature reserves are the last safe haven for many of the county's wild species of plants and animals. If it hadn't been for the foresight and actions of that small group of young people who set up the Trust in 1948, much more of Lincolnshire's wildlife would have been lost.

But it is not a time to become complacent. Just as more and more people are supporting the conservation movement, so there are others who destroy our countryside, often for little more than short-term profit, or through sheer ignorance.

Please use this book with enthusiasm. Visit the Trust nature reserves and experience these wild places for yourself. Take friends and family. No matter what the season, you will rediscover that wonderful feeling of nature in harmony and may well emerge back into your busy world refreshed and revitalised. Then, if you are not already a member, join the Trust and support its valuable work.

Professor David Bellamy
President of the Wildlife Trusts

Preface

This new edition of the Trust's *Nature Reserves Handbook* has been completely revised and includes fourteen new sites. Improved maps and a clearer indication of special facilities, disabled access and other features have been incorporated. Two sites that appeared in the previous edition are omitted. Bird's Wood, near Haxey, has been closed while sand extraction proceeds, which, under special arrangements negotiated with Hepworth Minerals plc, will eventually lead to the creation of a much larger wetland nature reserve. Wrawby Moor, near Elsham, is no longer managed by the Trust but continues to be cared for by the owners.

The information in this handbook has been compiled by many volunteers and members of the Trust's staff whose help is gratefully acknowledged. Special thanks are due to Clare Fry who assisted with editing and organised production, to Paul Miller for the sketches, and to John Redshaw who assisted with the maps.

Many of the nature reserves described in this handbook have been purchased with assistance from numerous donors and grant-aiding bodies. We cannot acknowledge them all individually but we record our thanks to all those who have helped. Where land has been received as a gift, this is acknowledged in the text.

We also record our gratitude to the following for generous financial support towards the costs of producing this book:

The Countryside Commission
Fenland Laundries Ltd

Stuart Crooks
1997

GEOFF TRINDER

Tortoiseshell Wood

About the Trust

The Lincolnshire Trust for Nature Conservation, formed in 1948, is a charity dedicated to safeguarding the countryside and wildlife of the historic county. It is one of a network of Wildlife Trusts that together form the largest voluntary organisation in the UK devoted to all aspects of wildlife protection. The Wildlife Trusts manage two-thirds of the country's nature reserves. The Lincolnshire Trust is the largest voluntary organisation in the county with over 9500 members. Its junior wing is the Wildlife Watch Club, which provides environmental activities for children aged between 8 and 15 years.

The activities and policies of the Trust are based on its extensive records of plants and animals, and of the wild places in Lincolnshire where they occur. In 1996 this data was used to compile a special report on the state of the county's wildlife habitats and on endangered species. Entitled *Nature in Lincolnshire: Towards a Biodiversity Strategy*, it aims to show how a wide range of organisations can contribute to the task of planning for a better deal for Lincolnshire's wildlife. Copies are available from bookshops or from the Trust's headquarters. The data is also made available to assist with planning and management in the countryside. Many hundreds of planning applications are screened and advice is given to local authorities, landowners and local communities. Education and interpretation are also important activities which depend on this resource.

However, it is the Trust's one hundred nature reserves that provide the most tangible and best known results of its work. They are enjoyed by perhaps half a million people each year.

To achieve its objectives the Trust relies heavily on the commitment and support of members. Much of the work is undertaken by volunteers with the support of field staff and a small administrative staff at the headquarters office. Many hundreds of volunteers work for the Trust in activities as diverse as manning sales points, managing nature reserves, conducting surveys of plants and animals, fundraising and checking planning lists.

Most of the Trust's income is provided by members and supporters through subscriptions, donations and legacies. If you enjoy visiting the nature reserves and care about them and the future of our countryside you are invited to support us by becoming a member and by remembering the Trust in your Will (see page 216). A membership brochure is available from visitor centres or the headquarters office.

About Nature Reserves

In order to understand their importance and the reasons for creating and protecting nature reserves it is helpful to know a little of the history of our countryside.

The countryside as we see it today has evolved from Man's use of the land over thousands of years. Around twelve thousand years ago Britain was still joined to the continent of Europe. As the glaciers of the last Ice Age retreated so wild plants and animals recolonised these islands, forming ecological associations determined by millions of years of evolution. When Neolithic man arrived, much of England was covered with deciduous woodland — the prehistoric wildwood. Since then, woodland has been cleared for agriculture, towns and industry. For over 6000 years the landscape changed relatively slowly, allowing wildlife to take advantage of the new conditions and to adapt to changed habitats. Many of our most attractive and valued habitats today have been created and maintained by traditional uses of the countryside. For instance, ancient woodland, which has links going back to the wildwood, has been modified by thousands of years of management for timber and fuel. Similarly, meadows and downland result from centuries of hay-making and grazing.

Since medieval times the pace of change has increased dramatically. By the early nineteenth century the last of Lincolnshire's great fenland areas had been drained and its downlands ploughed and enclosed. During the past 50 years developments in farming technology have removed most of the old meadows and hedgerows, and the proportion of permanent grassland has been drastically reduced in favour of intensively managed arable land. Other significant impacts have been the intensification of river and drainage management and the use of pesticides — organochlorine insecticides caused untold damage in the 1950s and 1960s. Agriculture, housing and industry now occupy most of the land surface; consequently, semi-natural areas have become scarce and the survival of wildlife has become increasingly threatened.

Nature reserves therefore offer havens for wildlife, where they may retreat and be given special care. Although wildlife continues to live on farmland and in urban areas it is no longer as varied and as widespread as it used to be. The richest communities of wildflowers, birds, insects and mammals in our county are now mostly to be found in nature reserves and in other specially protected areas. Nature reserves are also important as reservoirs from which wildlife can recolonise the surrounding countryside if conditions become favourable. Furthermore, they are valuable places for scientific study of wildlife and for the public to visit and enjoy. As well as enjoying the reserves it is hoped that visitors will come to appreciate and understand more about wildlife and be encouraged to help protect it.

Acquiring Nature Reserves

Many reserves are managed by the Trust under lease or agreements with the owners or tenants of the land. However, the freehold of the majority has been purchased by the Trust with funds from numerous sources — individual donations, appeals, grants and legacies. Legacies are especially important to the Trust and a special leaflet is available about how to leave a legacy in your Will for our work.

Managing Nature Reserves

All nature reserves require management. There are very few areas that can be left alone completely and many wildlife habitats require intensive management. There are several reasons for this. First, we have very little land in the British Isles that can be set aside solely for wildlife, so we have to make the most of what is available. Second, habitats have become fragmented and no longer have the natural variety formerly created by large herbivores and fire. Third, without intervention (by natural or artificial means) all habitats tend towards woodland through a process known as succession. If we are to continue to enjoy the earlier stages of succession, such as meadows and heathland, we must maintain management input to halt the successional process. In addition, we must manage nature reserves to facilitate access so that people may enjoy their visit without harming what they come to see. The most appropriate way of caring for many nature reserves is to return to traditional techniques, such as coppicing of woodland, managing meadows for hay and aftermath grazing, and extensive grazing systems on chalk and limestone grasslands.

Hebridean sheep

Grassland: Meadows

The old methods of farming grassland created conditions suited to a wide variety of wildflowers and associated insect species. Meadows were cut for hay and the re-growth, or 'aftermath', grazed later in the year. Since 1938 more than 99% of Lincolnshire's hay meadows have been converted to arable or to improved grassland. Farming improvements, such as re-seeding, drainage and fertilising, encourage coarser species, which become dominant, leaving no opportunities for the finer herbs and grasses. The Trust manages its hay meadow reserves by taking a late hay crop in July or August. The removal of the hay helps to keep fertility low and of course no fertilisers are used. Sheep or cattle are brought in from September onwards.

Meadows are at their best from May onwards until the hay is ready for cutting in mid-July.

Examples of hay meadows are: Heath's Meadows, Sotby Meadows, Kingerby Beck Meadows, Rush Furlong, Sedge Hole Close, Silverines Meadows, Hatton Meadows and the meadows adjoining woods — Bottleneck and Jackson's Meadows at Lawn Wood, and Porter's Lodge Meadows at Tortoiseshell Wood.

Grassland: Chalk and Limestone

Flower-rich grassland occurs on the chalk and limestone where there is a thin calcareous soil. Most of these areas were traditionally grazed by sheep. In order to maintain the floristic and invertebrate diversity of these sites the Trust either lets out the land for sheep grazing or, where this is not practicable, arranges for mowing and removal of the cuttings.

BARRIE WILKINSON

Hatton Meadows

BARRIE WILKINSON

Green-winged orchids, Heath's Meadows

The best time for seeing the wildflowers of chalk and limestone is from early to late summer.

Examples on the chalk are: Red Hill and Mill Hill Quarry, and on the limestone Duke's Covert and Copper Hill, Robert's Field and Ancaster Valley.

Woodland

The richest wildlife woods are those that have a continuous wooded history, perhaps with links going back to the original wildwood that covered much of England after the last Ice Age. Traditional management, which reached a peak of sophistication in the medieval period, was to grow large widely-spaced timber trees, with an understorey of smaller trees, which were cut on rotation to produce small wood for tools, fuel, poles and charcoal. This practice produces great variety, in some ways mimicking the conditions that would have existed in the wildwood. Many birds, for example nightingale, find a particular stage of the coppice cycle to their liking, moving from year to year to nest in the right stage of coppice re-growth. Other birds and many insects are adapted to woodland edge conditions, so that young coppice, glades and rides are important to them. The Trust is restoring traditional coppice working to many of its woods.

Examples of coppiced woods are: Tortoiseshell, Hoplands, Muckton, Goslings Corner and Rigsby.

Other woods, known as secondary woodland, have grown up on land that was previously heathland or farmland. Although generally less rich than ancient (or primary) woodland, they nevertheless have special characteristics and support a different range of species.

Examples of secondary woodland include: Roughton Moor Wood, Langholme Wood, Friskney Decoy Wood and the wooded parts of Moor Farm and Kirkby Moor.

A special type of woodland, known as carr woodland, occurs in the small valleys of the southern Wolds, on the Spilsby Sandstone. These are wet woodlands, which are dominated by ash and alder, and have a specialised range of plants, including golden-saxifrage and kingcup.

Examples of carr woodland are: Keal Carr, Lilley's Carr and Upper Sow Dale.

Heathland

Heaths are usually found on poor acidic soils and have characteristic grasses, heathers and other wildflowers. In Lincolnshire the majority of heaths are on the sandy soils of the Fen Edge around Woodhall Spa, the Trent Valley and the Cover Sands of north-west Lincolnshire. Heathland once occupied large tracts of the county, covering around 5000 acres until the 1920s, but forestry and agricultural reclamation have since destroyed the majority of this. Probably less than 600 acres now survive and much of that has suffered from increased drainage of surrounding land.

Examples of heathland are: Scotton Common, Linwood Warren, Kirkby Moor and Moor Farm.

Peatland

Poorly drained soils under certain conditions develop a special bog vegetation that is often dominated by *Sphagnum* mosses, which over long periods produce great depths of peat. There are few such places in lowland England but the great area of the Humberhead levels extends into north-west Lincolnshire. These raised bogs have great interest for their variety of acid-loving plants, such as sundew and bog rosemary. In some areas the vegetation is similar to heathland, with an abundance of heathers — ling, cross-leaved heath and bell heather. Most bogs in Britain have been drained and cut over for peat but some good areas remain in Lincolnshire and are being managed to restore water-levels and to control invading scrub.

Bog areas managed by the Trust are all in the Isle of Axholme: Crowle Moor (now part of the Humberhead Peatlands National Nature Reserve), Haxey Turbary and Epworth Turbary.

Wetlands

Fen, reedbed and open water habitats in Lincolnshire are almost all artificial in origin, but they may be regarded as substitutes for the great fenland areas drained in the 18th and 19th centuries. Some wetlands arose as a result of industrial activities, such as tile and brick making, and sand and gravel extraction. Others were borrow-pits used in the construction of sea banks and railways. Still others were formed as washlands to protect farmland from flooding. Today these places offer a great variety of habitats for aquatic wildlife, ranging from reed warblers and bearded tits — found almost exclusively in reedbeds — to grebes, ducks and wading birds.

Examples of open water are: Boultham Mere and Fairfield Pit. Reedbeds are situated at Far Ings, Pasture Wharf, Barrow Haven, Huttoft Bank Pit and Chapel Pit. Whisby Nature Park, Messingham Sand Quarry, Burton Gravel Pits and Kirkby Gravel Pit all result from sand and gravel extraction. Washlands are to be found at Baston Fen and Thurlby Fen Slipe.

Coastland

Of all Lincolnshire's wild places, it is the coast that is the most nearly natural. Although reclamation, sea defences and industry have altered large stretches, there is a great deal that remains unspoilt. The whole of the Wash is one of England's finest wildlife areas, especially noted for wading birds and wildfowl which inhabit the sand and mudflats. Between Grimsby and Gibraltar Point there are extensive sand dunes, clothed in dune grassland and scrub of mainly sea buckthorn. There are also saltmarshes, freshwater slacks, and seashore of shingle and sandy beaches.

The processes of accretion and erosion, and tides and storms make the coast a dynamic place that is forever changing. It is the ability of plants and animals to adapt to these shifting conditions that makes the coastland so rich and interesting. From the first colonisers of the bare mud, through saltmarsh, to mobile and stable dunes, and damp hollows and open shore, the coast is a kaleidoscope of habitats.

Examples of coastal nature reserves are: Gibraltar Point, Saltfleetby–Theddlethorpe Dunes, Donna Nook and Frampton Marsh.

Other Habitats

As well as the characteristic habitats described above, there are many other habitats in the county, almost all of which occur on Trust nature reserves. Rivers provide a special environment, some of which is protected in reserves. Examples of rivers are: the River Bain at Kirkby Gravel Pit and the River Glen at Baston Fen. Hedgerows are best represented at Heath's Meadows, Sotby Meadows, Kingerby Beck Meadows and Hatton Meadows. Disused railway lines with scrub and grassland make fine wildlife sanctuaries. Examples of railway lines are: Horbling, Axholme (Belton and Haxey), Willoughby Branch and Woolsthorpe Lines, and the Witham Way.

SSSI

Sites of Special Scientific Interest are notified by the government's conservation agency, English Nature. They are selected as the best and most representative examples of wildlife habitats in the country. Many Trust nature reserves have this status, which is indicated at the head of each entry.

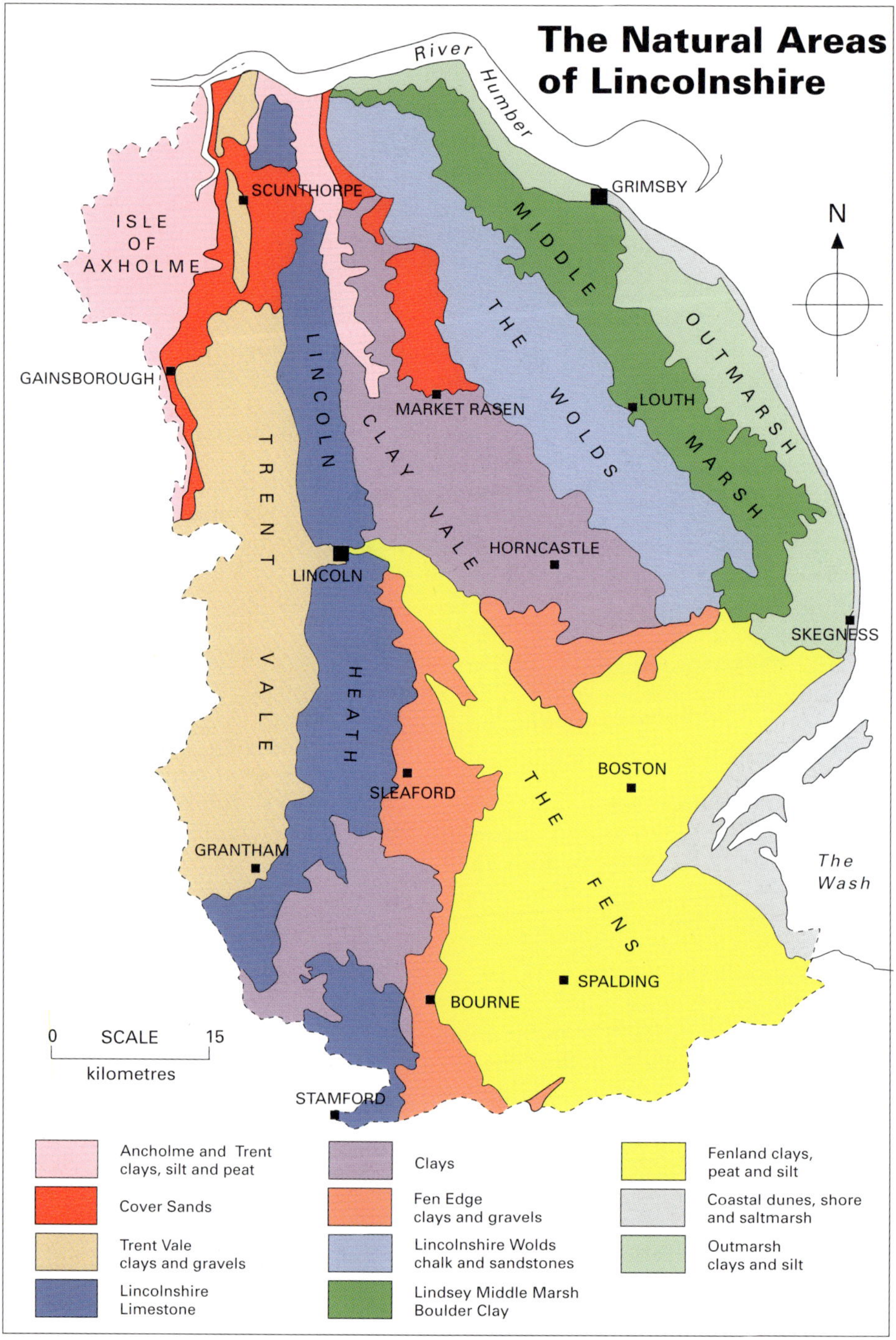
The Natural Areas of Lincolnshire
River Humber
GRIMSBY
SCUNTHORPE
ISLE OF AXHOLME
N
MIDDLE MARSH
THE WOLDS
OUTMARSH
GAINSBOROUGH
LINCOLN CLAY VALE
MARKET RASEN
LOUTH
TRENT VALE
HORNCASTLE
LINCOLN
SKEGNESS
HEATH
BOSTON
SLEAFORD
THE FENS
GRANTHAM
The Wash
SPALDING
BOURNE
0 SCALE 15
kilometres
STAMFORD
Ancholme and Trent clays, silt and peat
Cover Sands
Trent Vale clays and gravels
Lincolnshire Limestone
Clays
Fen Edge clays and gravels
Lincolnshire Wolds chalk and sandstones
Lindsey Middle Marsh Boulder Clay
Fenland clays, peat and silt
Coastal dunes, shore and saltmarsh
Outmarsh clays and silt

Geology and Natural Areas

The underlying solid geology of Lincolnshire is of alternate hard and soft rocks dipping gently eastwards. However, much is covered by glacial boulder clay and gravels, blown sand, peat and alluvium, all of which influence soils and drainage and, in turn, the variety of wildlife habitats. Heathlands occur on the acid soils of sand and peat, limestone and chalk grassland only where those rocks outcrop, and wildflower-rich meadows on the clays.

There are two lines of hills — the limestone Heath and the mainly chalk Wolds, between which is a clay vale broadening southwards to the peat and silt-filled depression of the Fens. To the west lie the Trent vale and the Isle of Axholme, while to the east is the Lincolnshire Marsh fringed by a sand-dune and saltmarsh coastline.

In the Isle of Axholme low hills separate peatlands and patchy sands on the west from the alluvial carrlands of the Trent. East of the Trent around Scunthorpe are former ironstone quarries and extensive cover sands supporting heathlands. To the west of Lincoln are much quarried sands and gravels with areas of open water, and the upper Witham and its tributary, the Brant, flow on clays.

The limestone ridge is narrow north of the Lincoln Gap and broadens southwards, with a distinctive west-facing scarp slope. South of the gravel-filled Ancaster gap the broad Kesteven uplands have patches of glacial boulder clay with remnants of natural woodland, narrow dissecting valleys and the legacy of ironstone quarrying.

The Lincolnshire clay vale is also narrow in the Ancholme carrlands, with cover sands to the south. In the centre of the county extensive boulder clay supports ancient woodlands of the Bardney Forest, while the heathy moors in the Woodhall area are on flinty gravels washed from the Wolds by glacial meltwaters.

The chalk-capped Wolds extend south-east from the Humber, with a west-facing scarp and in the east unusual valleys cut by glacial meltwaters, while the eastern margin is truncated by a line of former chalk sea cliffs. In the south-west is a wide ledge of Spilsby sandstone where streams have cut through to the underlying clay, creating a very different kind of landscape.

The Lincolnshire Marsh is divided into the undulating boulder clay Middle Marsh and the flat Outmarsh of marine silts, the latter exploited for brick and tile making along the Humber bank. Along the edge of the Middle Marsh from Barton to Tetney are artesian blow wells where lenses of sand allow water from the chalk below to reach the surface.

Of the once large area of poorly drained fens, hardly a vestige remains — only washlands, decoy woods and drain-sides provide something of a natural habitat. However, where there has been quarrying of the fen-edge gravels in the south-west, open water habitats have been created.

Lincolnshire's long coastline — over 100 miles from the Trent to the Nene — includes the great estuaries of the Humber and the Wash, the latter with wide fringing saltmarshes.

Distribution of Reserves

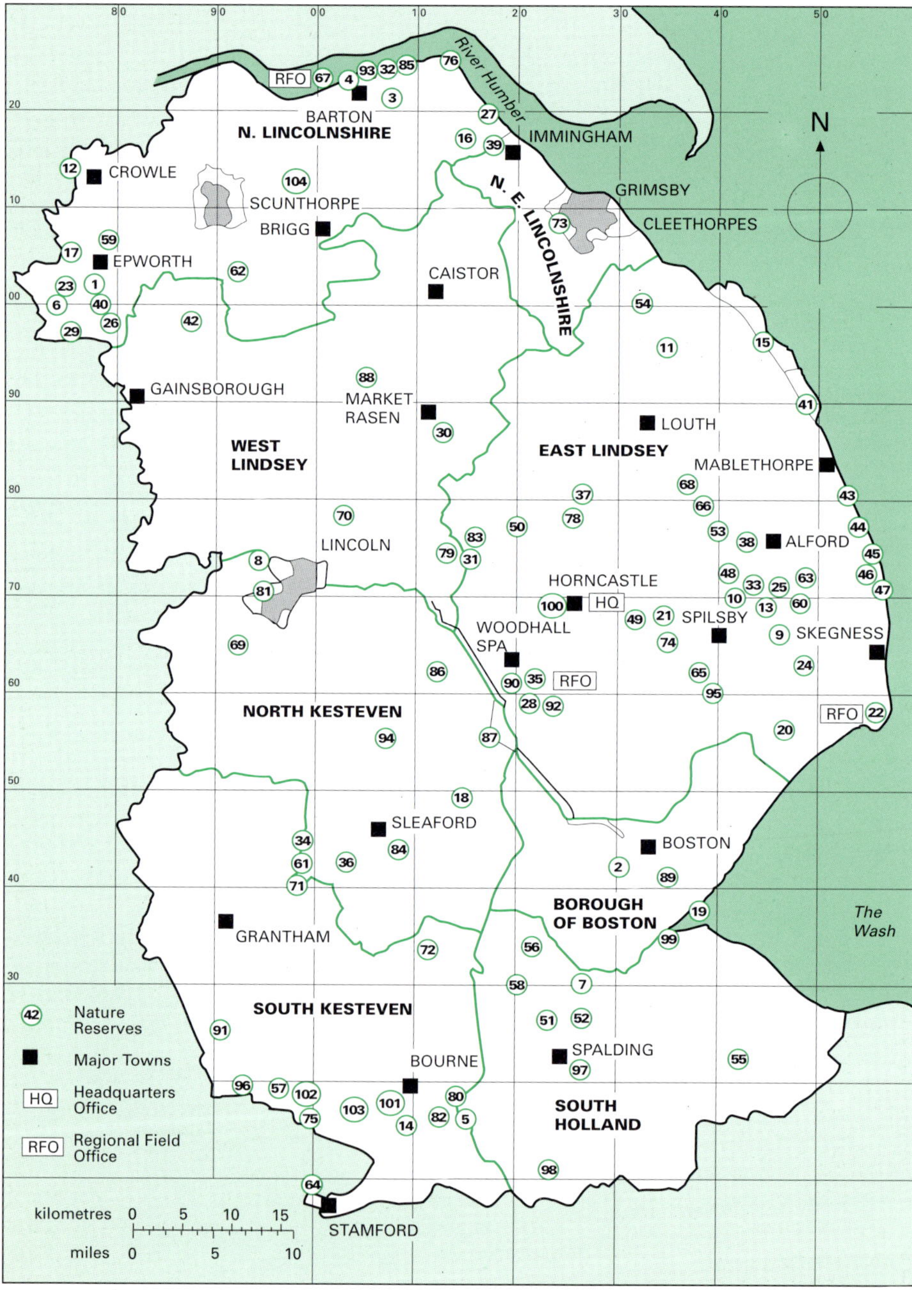

List of Reserves

Reserves and Education

The Trust's reserves represent an enormous resource for educational purposes. Many are suitable for class visits and most offer opportunities for individual or small group project work.

At Far Ings, Whisby Nature Park and Gibraltar Point there are special facilities, including laboratories or field study rooms, classrooms, interpretive displays and leaflets. At these, and at Snipe Dales, staff support can be provided. Enquiries should be addressed to the Warden of the individual site or to the headquarters at Banovallum House. At Gibraltar Point there is also residential accommodation suitable for schools use.

For specialist studies advice may be obtained from Trust staff. In general groups are expected to book visits in advance.

CLIFFORD JUKES

Helping to identify grasshoppers, Kirkby Moor

Access to the Reserves

Information about access to each reserve is indicated by symbols under the reserve entries. All the nature reserves showing the symbol 🚶 are open to the public. A few reserves have restricted access and these are indicated by the symbol **S**, but most of these may be visited with a member's permit at any time. Permits for group visits (10 or more) and for educational use may also be obtained from the Trust's headquarters. Membership of the Trust provides an annual permit for visiting almost all the reserves.

Disabled Access

The wheelchair symbol shown for some reserves indicates that disabled access is possible, at least in good weather conditions. It does not imply that there are special facilities. Good all-weather wheelchair access is available at the four major visitor sites — Whisby Nature Park, Gibraltar Point, Snipe Dales and Far Ings. Reserves marked with a wheelchair and key symbol indicate that a RADAR key will open a locked gate.

In addition to the disabled access indicated by the symbols, there are several sites at which good views can be had from car parks or verges. These include Red Hill, Gibraltar Point, Whisby Nature Park, Copper Hill (part of Duke's Covert), South Witham Verges and Fairfield Pit.

DAVID BROMWICH

Whisby Nature Park

Visiting the Reserves

Nature reserves are special places that are carefully and sensitively managed for their wildlife. In order to help us to look after them please observe the following points:

1 No collecting
Please do not remove any plants or animals. For teaching and serious study, collecting permits may be issued by the Trust's headquarters.

2 No tents, hides or vehicles
Tents and hides are not allowed except by special permission. Vehicles may not be taken into reserves except where special parking provision is made.

3 Go quietly and follow the Country Code
Please go quietly and observe the Country Code. Smoking and the lighting of fires and stoves are strictly prohibited. Please take your litter home.

4 Dogs
Dogs are generally not welcome. At a few sites identified by a symbol, dogs are allowed provided they are kept on a lead. Please remember to clean up after your dog.

5 Special restrictions
Special restrictions or conditions may apply at some of the reserves. Some of these are indicated in the text but please look out for notices posted at entrances and elsewhere. Special care is needed when visiting meadows before haytime.

6 Waymarked routes
Where there are waymarked routes please follow the paths. The routes are liable to change from time to time. Visitors are expected to comply with any proper request of a Warden or other Trust officer.

WARNING: Cars parked at nature reserves are increasingly vulnerable to theft and damage. Do not leave valuables in the car and report any persons behaving suspiciously to the police.

Records and observations from the nature reserves are always welcomed by the Reserve Managers or the Trust's headquarters.

The Trust accepts no responsibility for any loss, injury or damage, howsoever caused, that may be sustained while visiting any nature reserve.

Take nothing but photographs
Leave nothing but footprints
Kill nothing but time

When to Visit the Reserves

Advice on when to visit reserves is not straightforward as it depends greatly on the type of reserve and its habitats. Although each reserve may have its annual highlights, almost all are worthwhile at any time of the year. However, it is possible to suggest some guidelines:

Meadows are at their best from May, with a peak of flowers on show in June, until hay time in mid-July.

Woods are best from April until mid-June, after which the developing canopy shades out the early flowers and makes birdwatching more difficult. The period between September and November is of course a good time for autumn colour and, in most seasons, is the best time for fungi.

Heathlands are at their best in high summer, from July through to September.

Wetlands are best for breeding birds and flowering plants in May and June. The spring and autumn are best for migratory birds, especially in areas of shallow water.

Coastlands have much to offer in all seasons: wildfowl and wading birds in winter; sea lavender in late July and early August; the famous marsh orchids at Saltfleetby in mid-June; and breeding birds from May to August.

How to Use This Guide

The nature reserves are arranged in alphabetical order. The map on page 16, with its accompanying list, may be helpful in locating reserves. The indexes on pages 210 and 214 should help to identify any reserves whose names you are unsure of.

At the head of each entry the following details are set out in a standard form:
Reserve Name
Parish
OS: Number of the Ordnance Survey 1:50000 map
GR: Grid reference, usually of entrance or of extremities for linear sites
Area in hectares followed by acres in parentheses
Status and date of establishment
Map reference (see page 17)

Symbols and Abbreviations

The following symbols are used to indicate special features and the main wildlife interest for each reserve:

- Open to the public
- Special restrictions or conditions (see details in text)
- Wheelchair access to at least part of the reserve
- RADAR key for disabled persons to open locked gate
- Dogs permitted on a lead
- Visitor centre or information point
- Leaflet available
- WC Toilets
- Birdwatching hide
- Waymarked route
- Of special interest to birdwatchers
- Interesting mammals or other vertebrates
- Interesting plants
- Interesting fungi
- Interesting insects or other invertebrates
- SSSI Site of Special Scientific Interest

Site Maps and Location Plans

Site maps show the location of the reserve, its boundaries, where to park and enter the reserve and the main footpaths and any hides. The following is a key to the abbreviations used for each map:

P Parking space
E Entrance
H Hide
Waymarked route

NATURE RESERVES

Ancaster Valley

Ancaster
OS: 130 GR: SK 984434 9.7 hectares (23.9 acres)
Freehold 1982
Map ref: 61

SSSI

Location and Access

The reserve is situated on the south side of the A153 Sleaford–Grantham road close to the village of Ancaster. The entrance is reached via a short length of bridleway signposted from the road, about 100 m to the east of the Ancaster crossroads. A public footpath runs through part of the valley. Access is by a stile. The gate is kept locked but access for disabled persons is possible using the RADAR lock. It is advisable to park cars in the village. From time to time there will be sheep grazing in the reserve.

Description and Management

This narrow, steep-sided valley is covered with limestone grassland, scrub and woodland. It is one of the finest sites for limestone flowers in the county. Recorded species include the rare and beautiful pasqueflower, bee and fragrant orchids, dyer's greenweed, dropwort, spring cinquefoil, rock-rose, spring-sedge and many others. There is an attractive woodland hanger, consisting chiefly of beech, on the eastern slope of the valley, which has yew, box, wayfaring tree, barberry and wild clematis on its edge, and woodland plants, such as violet and early-purple orchid. Large numbers of butterflies and bumble-bees may be seen on calm sunny days. The wide variety of birds include green and great spotted woodpeckers.

Large areas of the steep banks had become overgrown with scrub of gorse and hawthorn in the 1970s, and, since the reserve was acquired in 1982, one of the main management tasks has been clearance and control of scrub in order to retain the flower-rich grassland. This is now managed with light sheep grazing. The woodland is being managed and planted with native species, such as oak, ash and field maple, where there are gaps.

The origin of Ancaster Valley is uncertain, but it is thought to be a glacial spillway formed at the close of the Ice Age when meltwater from the ice cap on higher ground poured into the Ancaster Gap — the former course of the River Trent before ice forced its course further north.

Small copper butterfly

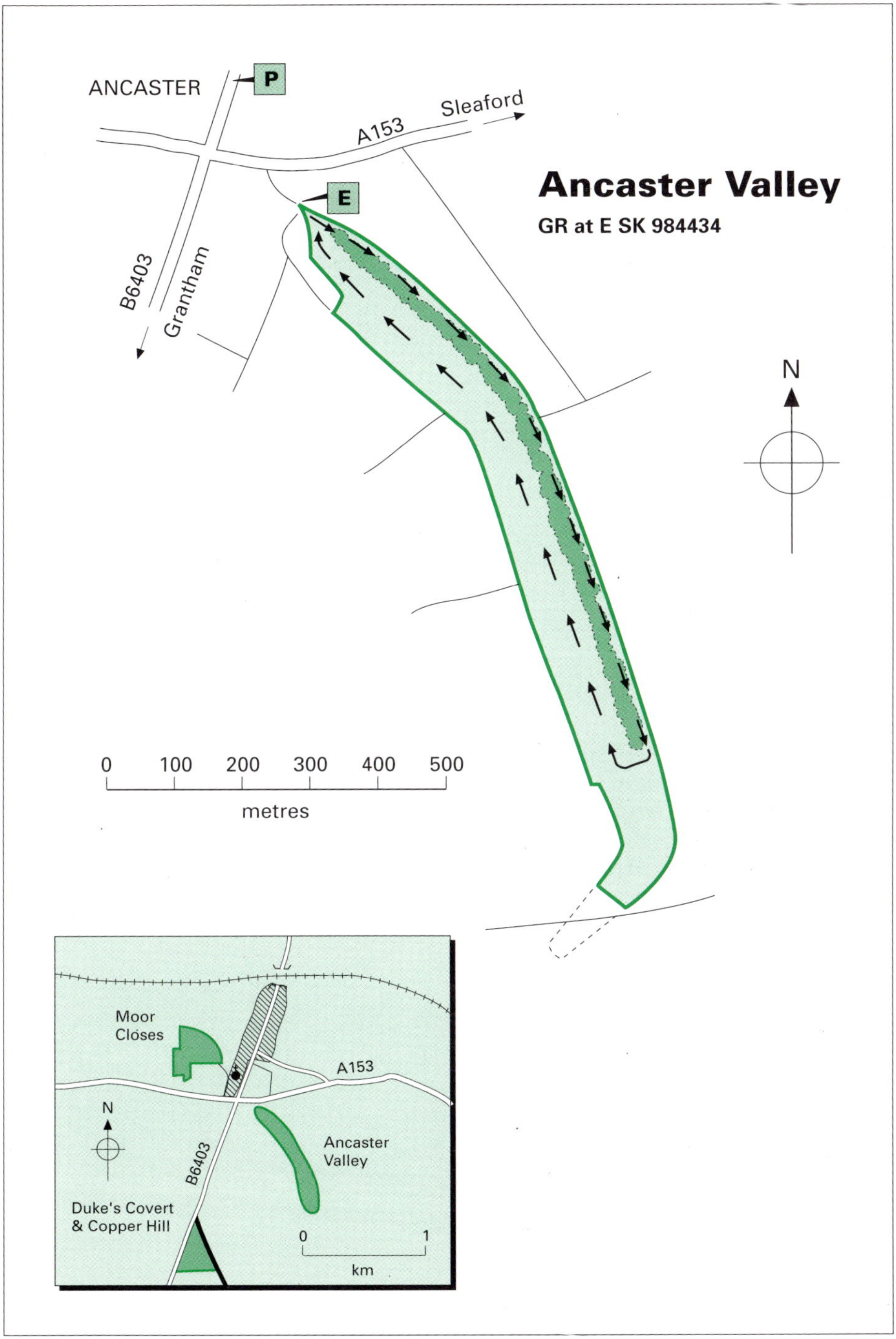
ANCASTER
P
A153
Sleaford
E
Ancaster Valley
GR at E SK 984434
B6403
Grantham
N
0
100
200
300
400
500
metres
Moor Closes
A153
N
Ancaster Valley
B6403
Duke's Covert & Copper Hill
0
1
km

Arnold's Meadow

Spalding
OS: 131 GR: TF 260216 2.6 hectares (6.5 acres)
The Arnold Smith Memorial Reserve. Freehold 1992
Map ref: 97

Location and Access

The meadow is situated on the east side of the Coronation Channel at Spalding. Access is reached from Peck's Drove, crossing a Bailey bridge over the channel, onto Childer's South Drove. There is parking space for a few cars on the grass verge. The entrance is through a small hand-gate. Turn left onto a brick path past a caravan and through another gate from where paths to both hides are indicated. Please close all the gates after entering and before leaving, and keep to the paths as the meadow is used for hay cropping.

Description and Management

Half of the area of this triangular-shaped reserve is taken up by a hay meadow, the centre of which is flooded in autumn and winter. The remaining half consists of three areas of water: a moat round an island and two shallow ponds overlooked by hides. The reserve is bounded by mature hedgerows and trees, with a small wooded area in the northern corner.

The variety of habitat attracts many birds: 73 species have been recorded, including kingfisher, and 35 species have been known to breed, including sparrowhawk and jay. Seventeen species of butterfly have also been recorded. There is a good population of water voles, toads, frogs and grass snakes. Plants are abundant on the reserve with 153 species recorded on the heavy clay. The wet area of the meadow has purple-loosestrife, brooklime, ragged-robin and butterbur.

Before 1968 the area was a bare field, and the reserve has therefore been entirely man-made. The meadow is managed as a traditional hay meadow by taking a late hay crop, while vegetation in the ponds is controlled in order to retain open water. The buildings are used by the Trust's local volunteer group that help manage reserves in south Lincolnshire.

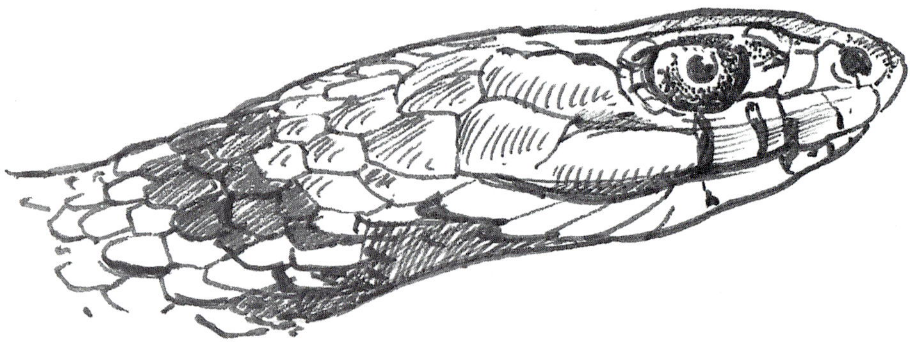

Grass snake

Arnold's Meadow

GR at E TF 260216

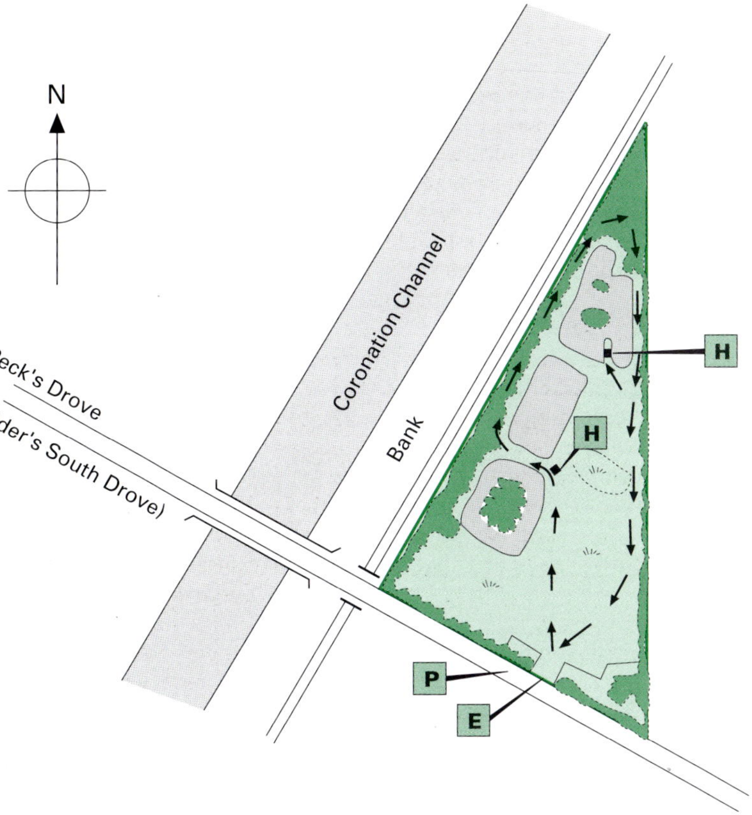

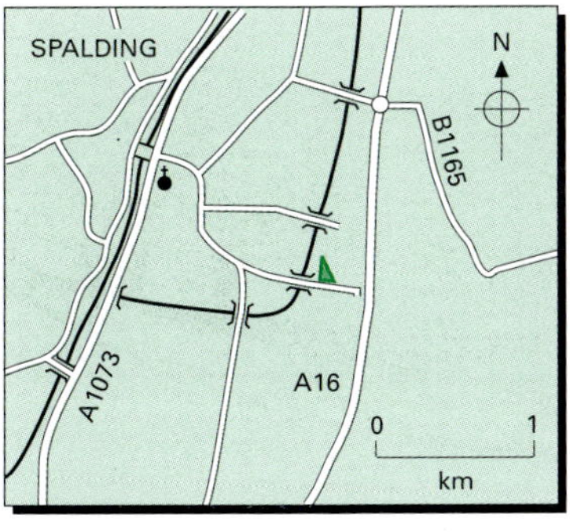

Axholme Line — Belton

Belton
OS: 112 GR: SE 782054 to 787068 4.8 hectares (12 acres)
Leasehold 1982
Map ref: 59

Location and Access

Situated between the village of Belton and the Belton picnic site, this length of disused railway line is reached either from Jeffrey Lane in Belton or from the east side of the A161 road opposite the picnic site. Another length of the same line — the former Axholme Light Railway — is managed by the Trust near Haxey (see page 30).

Description and Management

The reserve consists of grassland and scrub with hawthorn, blackthorn, ash, dog rose, privet and hop. There is a good grassland flora with creeping cinquefoil, ribbed melilot, meadow vetchling, flixweed, common knapweed and perforate St John's-wort. Butterflies include ringlet and meadow brown. Among nesting birds are whitethroat, dunnock and linnet.

Management includes the control of invading scrub, annual mowing of grassland and more frequent mowing of the track bed.

Perforate St John's-wort

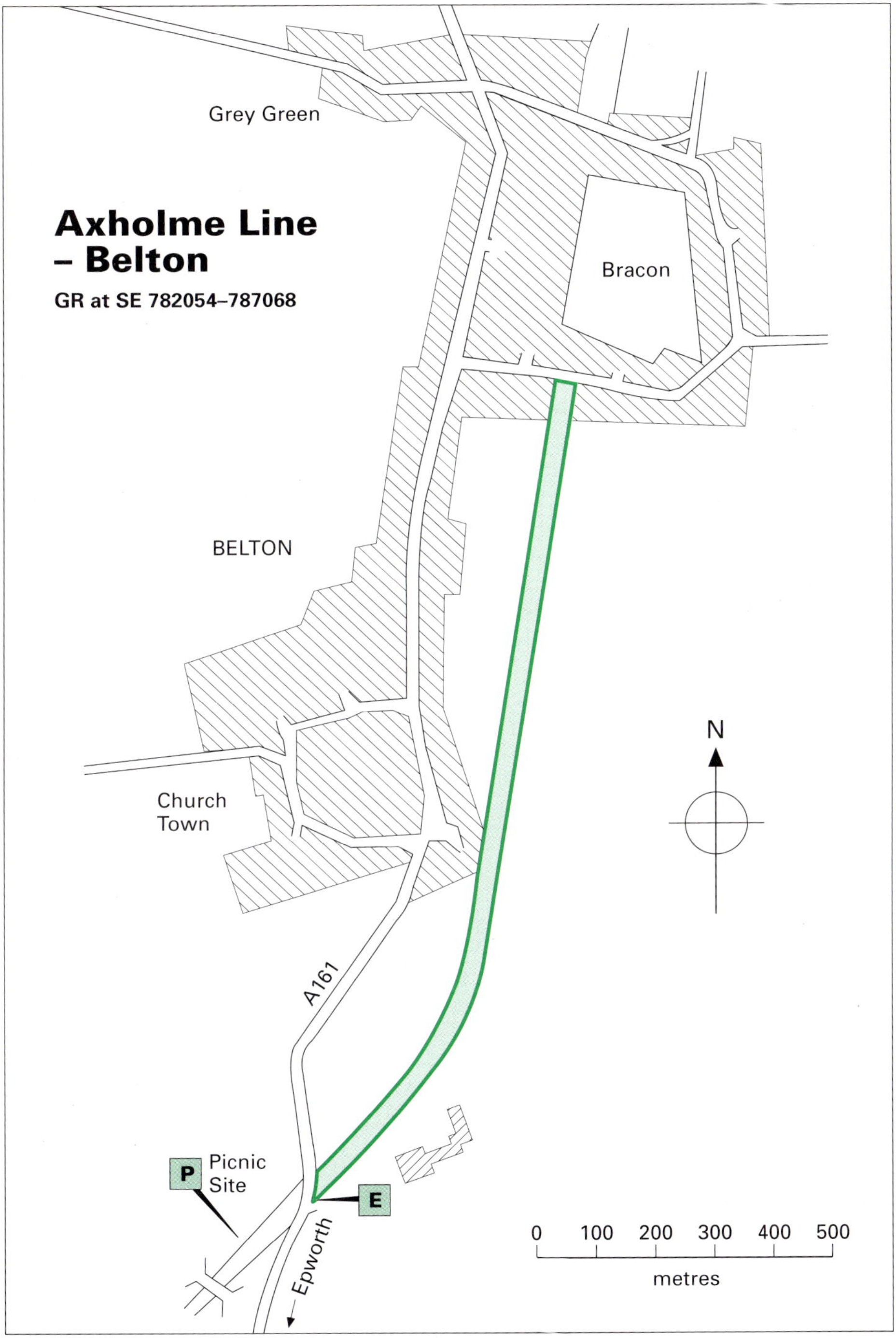

Axholme Line
– Belton
GR at SE 782054–787068
Grey Green
Bracon
BELTON
Church
Town
N
A161
Picnic
Site
P
E
Epworth
0
100
200
300
400
500
metres

Axholme Line — Haxey

Haxey
OS: 112 GR: SE 773000 to 771024 13.7 hectares (34 acres)
Leasehold 1978
Map ref: 1

Location and Access

Situated to the north of Haxey in the Isle of Axholme, the reserve consists of the disused former Axholme Light Railway running north–south, parallel to and west of the A161 road. The main access is from Blackmoor Road. The line runs from this point for some 2.5 km (1.5 miles) northwards. A car park is available close to the junction of Blackmoor Road and Marlborough Road in Haxey. Other access points exist off Nooking Road and at Burnham Hill but cars should not be parked where they will obstruct farm or house access tracks. The main walk along the bed of the track is open to the public. Visitors may also examine the embankments but care should be taken to avoid damage to the friable marl and gypsum. The reserve provides a level walking route especially suitable for the elderly and disabled. Another length of the line north from the Belton picnic site is also managed by the Trust (see page 28).

Description and Management

The southernmost mile of the line is in a cutting, while the remaining northern section is embanked. The cutting passes through the base-rich Keuper Marl, which supports a diverse flora. As well as hawthorn the slopes carry ash, aspen, dog rose, elder and oak. Field maple and agrimony occur here, the latter at its only location in the Isle of Axholme. The speciality of the reserve, the green-winged orchid, grows in considerable numbers both on the banks and in the ballast. Eighteen butterfly species have been recorded, including the dingy skipper, though the latter has not been seen for some years and is now thought to be extinct on the site. The area is also good for birds, especially long-tailed tit, bullfinch, redwing in winter, and whitethroat and willow warbler in summer. Barn and long-eared owls hunt the area. From the embanked part of the line good views over the Isle are obtained; two other Trust reserves can also be seen: Haxey and Epworth Turbaries.

Much of the line had become engulfed by dense hawthorn scrub, and one of the main management tasks undertaken by the Trust since acquiring the reserve in 1978 has been scrub clearance in order to retain the interesting grassland.

Dingy skipper butterfly

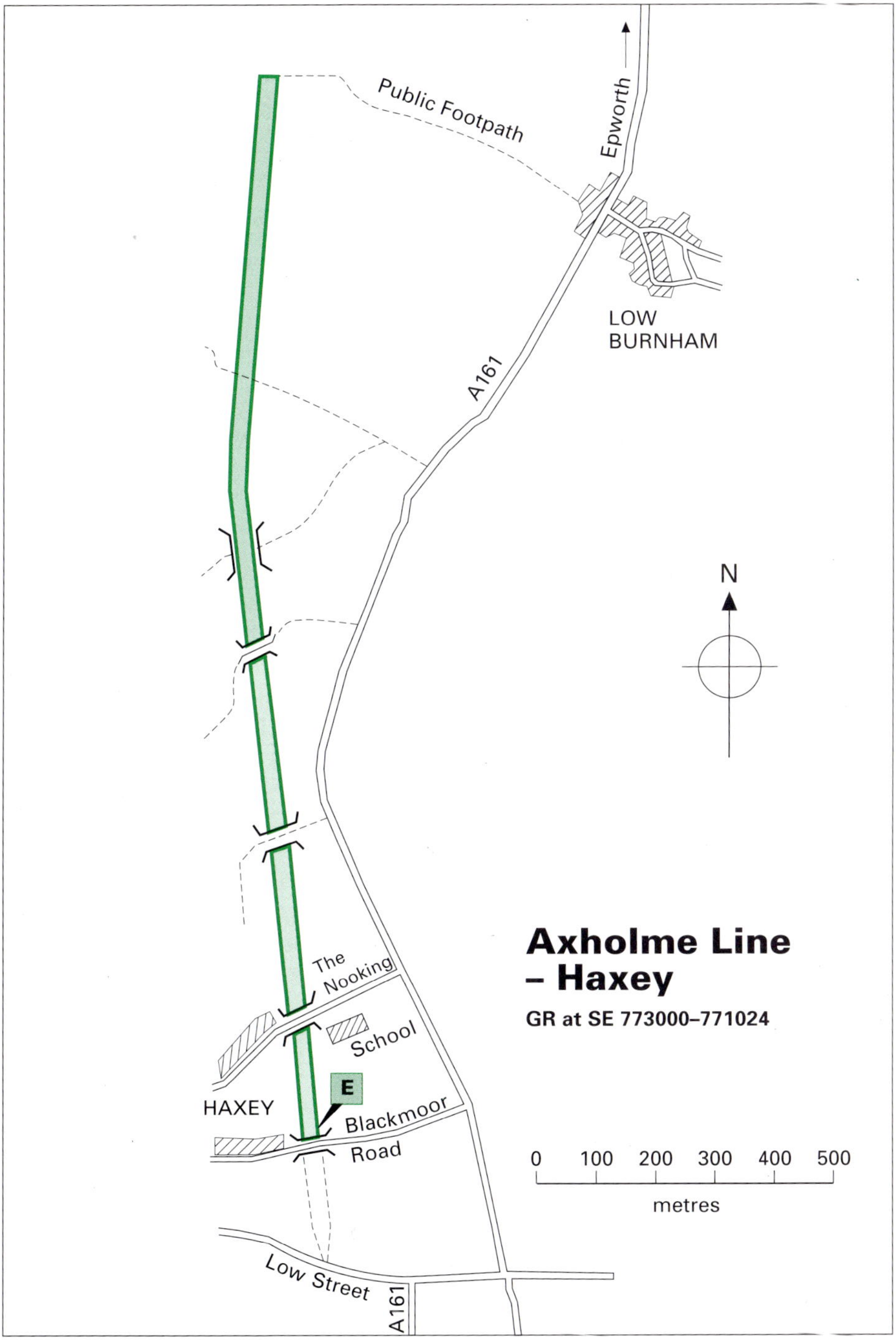
Public Footpath
Epworth
LOW BURNHAM
A161
N
The Nooking
School
E
HAXEY
Blackmoor Road
Low Street
A161
Axholme Line – Haxey
GR at SE 773000–771024
0 100 200 300 400 500
metres

Banovallum House

Horncastle
OS: 122   GR: TF 257695   1.5 hectares (3.7 acres)
Freehold 1993
Map ref: 100

Location and Access

Banovallum House is the Trust's headquarters, and is situated a short distance from the Horncastle Market Place in Manor House Street. Banovallum House was built in 1790 and was extended in about 1850. A leaflet on the history of the house is available from Reception. Limited parking is available during office hours. Visitors are requested to make themselves known at Reception before entering the grounds.

Description and Management

The grounds of Banovallum House are situated between the Jubilee Way and the River Bain, and include the North Staunch Field, a wildlife garden, a walled garden and about 200 m of river bank. The grounds are open to visitors during office hours. A small shop selling gift items is situated in the Reception area. The grounds are good for a variety of garden birds, including finches and tits, spotted flycatcher, song thrush and pied wagtail. Kingfishers can be seen along the river.

Waymarked Route 0.5 km (0.3 miles)

The route starts on the eastern side of the house, nearest the Market Place. Cross the stile into the meadow. The path follows the right-hand boundary of the meadow to the river bank. At certain times of the year a few of the Trust's Hebridean sheep may be seen here. A hay crop is usually taken in late July. At the river turn left along the bank. The river is quite clean and is rich in aquatic life, including fish such as chub, trout and bream. The path continues along the bank and passes into a small copse of willows and other newly planted trees. The largest oak tree is known as the Cranbrook Oak, having been planted by Lord Cranbrook at the opening ceremony for the headquarters in 1993. The path continues into the North Staunch Field, a narrow strip of pasture between the river and the road. There are some fine old willows overhanging the river bank. Bear left away from the bank and head back across the field towards the gateway near the road. A gate leads into the wildlife garden, which is still under development. Some of the features to note are the small woodland, the green lane access to the meadow, and the pond. Other habitats are being created in order to demonstrate what can be done to encourage garden wildlife and to help people to plan small conservation areas. Plants that have been introduced to the pond and its margins include greater spearwort, bogbean, water-soldier, yellow flag and purple-loosestrife.

Follow the path to the arch in the brick wall. The archway leads into the walled garden, which is being restored on traditional lines, with wide borders. You are welcome to stroll around the garden. Many of the plants are labelled. On a mound next to the roadside wall there is a fine old walnut, which is reputed to have been planted by Sir Joseph Banks. The path from the walled garden leads back to the car park.

Banovallum House

GR at E TF 257695

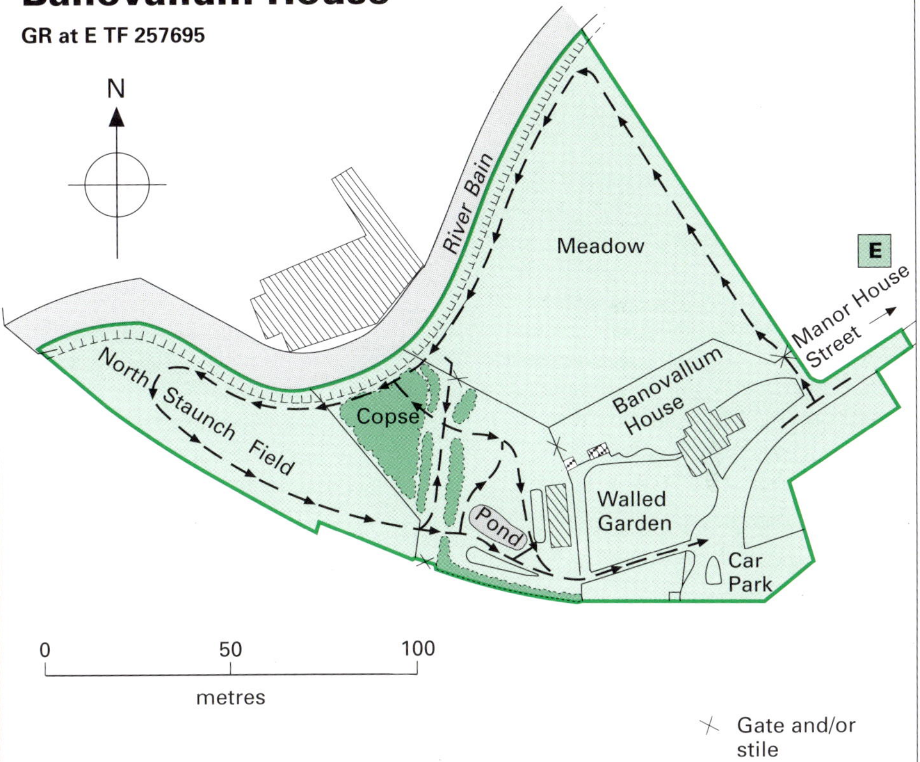

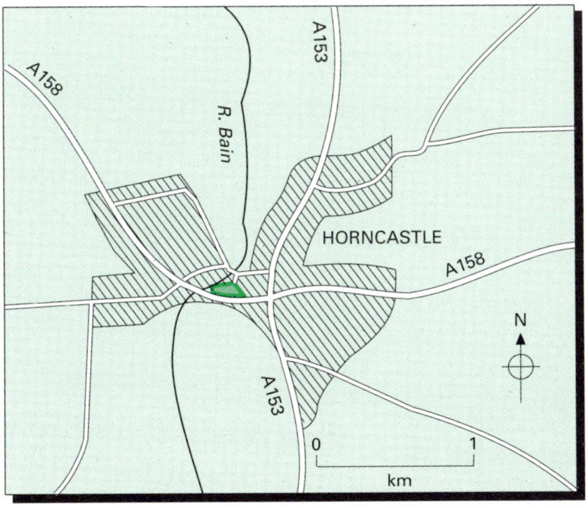

Baptist Cemetery

Boston
OS: 131 GR: TF 312434 0.28 hectares (0.7 acres)
Management agreement 1971
Map ref: 2

S

Location and Access

Turning off the Sleaford road (A52) on to South Parade, the cemetery is situated on the right, and is about 1.5 km (0.9 miles) from the centre of Boston. Access is through a private garden: please contact the headquarters office for details.

Description and Management

This disused cemetery in the residential outskirts of Boston provides an interesting habitat for a variety of birds. There are some large standard trees and sallows. Old overgrown hawthorns on the southern boundary have been coppiced in order to encourage a greater variety of wildflowers. A pond has been created for feeding birds and spawning frogs.

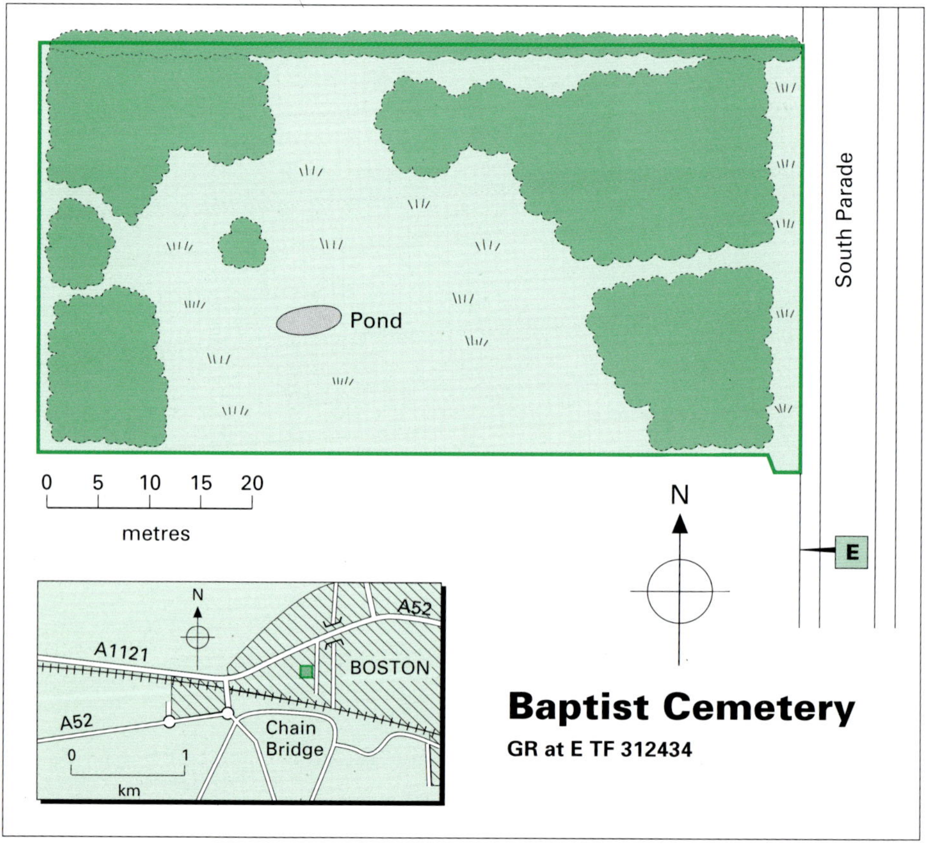

Barrow Blow Wells

Barrow-upon-Humber
OS: 112 GR: TA 073221 2.7 hectares (6.8 acres)
Freehold 1968. Gift from Mr JFD Hargreaves
Map ref: 3

Location and Access

The reserve is located on the north side of Barrow-upon-Humber and is reached along a track leading northwards from beside 'Open Holme' in Cherry Lane. Visitors are requested to keep to the waymarked route.

Description and Management

The reserve consists of reedmarsh and woodland with two blow wells, which are natural artesian springs caused by water under pressure escaping from the chalk through the overlying boulder clay. Such springs were once numerous in the Lincolnshire Marsh but most have dried up due to the depression of water-tables by increased abstraction. The larger of the two wells forms a pool full of clear water in which grows a mass of the pale-green, common water-starwort. Around the wells is a marsh, which is mainly colonised by reed sweet-grass. The wood may once have been an old osier bed. Beyond the wood is an area of wet pasture dominated by sedges. The Barrow Beck flows along the northern boundary. There is a variety of birds, including sedge warbler. Kingfishers are seen occasionally.

Management consists of maintaining fences and paths, keeping the wells clear of fallen timber and replanting woodland.

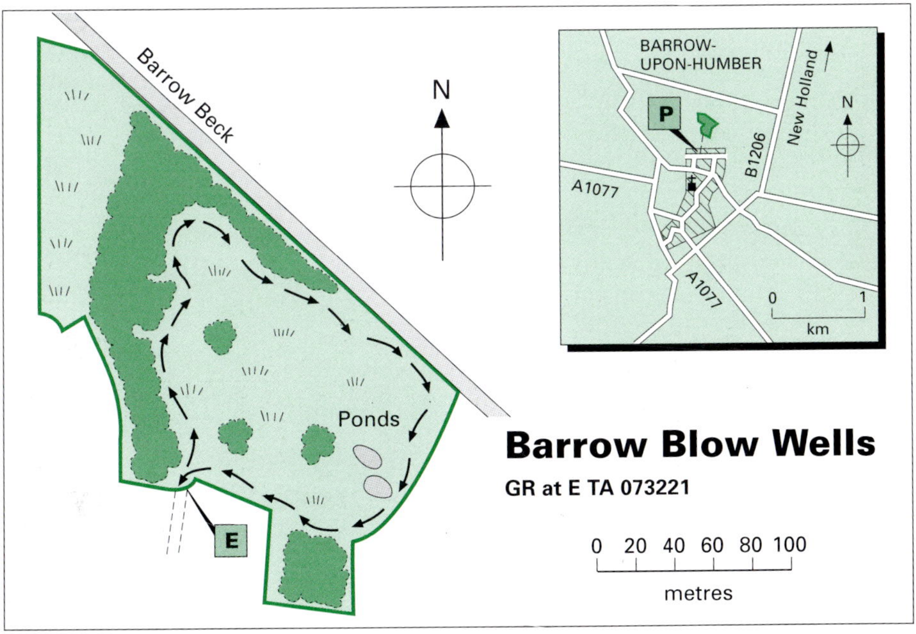

Barrow Haven Reedbed

Barrow-upon-Humber
OS: 112 GR: TA 061236 12.7 hectares (31.6 acres)
Freehold 1989
Map ref: 32

SSSI

Location and Access

The reserve is bounded by the New Holland–Barton railway to the south, the Humber estuary to the north, Barrow Haven to the east and a private access road to the west. Access is on foot across the bridge from the east side of Barrow Haven, which is accessible by a public road and where there is a small car park. A public footpath runs to the north of the clay pits and cuts through the reserve. For safety reasons and in order to avoid disturbance visitors are requested to keep to established or marked tracks. A viewing hide is reached from the path in the south-east corner.

Description and Management

One of the most important of the series of disused flooded clay pits on the Humber Bank. Clay digging was completed here in about 1950, leaving a pit of varying water depth. Extensive reedbeds (*Phragmites*, known sometimes as Norfolk Reed) and stands of great reedmace with areas of open water have since developed. The aquatic flora here is richer than in many of the other Humber Bank pits, extensive beds of mare's-tail being a notable feature. Patches of hawthorn scrub, bramble thicket and rough grassland with a pleasantly varied flora surround the water and reedbed areas. The reserve includes the sea bank and a strip of saltmarsh extending to mean high water mark. More than 150 species of birds have been recorded of which 60 have been known to breed, including great crested grebe, pochard, tufted duck, ruddy duck, water rail, bearded tit, and reed and sedge warblers. Bitterns bred until the 1960s and still occur occasionally as visitors in winter.

Management includes scrub control, reed-cutting and the maintenance of shallow margins for the benefit of reedbed birds.

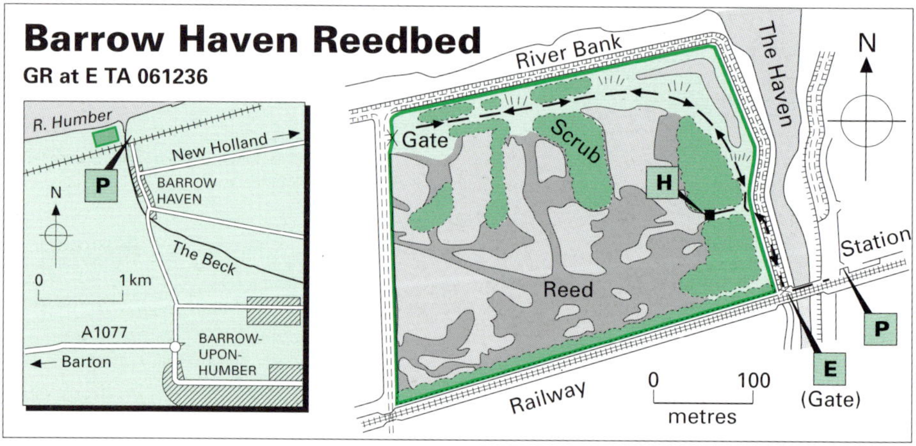

Baston Fen

Baston
OS: 130 GR: TF 145176 33 hectares (81.5 acres)
Freehold 1967
Map ref: 5

SSSI

Location and Access

The reserve is situated between the River Glen and the Counter Drain. The entrance is at the east end of the reserve about 1 km (0.6 miles) south-west of the junction of the Baston–Pode Hole and Bourne–Tongue End roads. Unless otherwise indicated, parking space is available over the concrete bridge and along the track to the left. Visitors are requested not to park on the track running straight from the bridge to the River Glen bank as this obstructs farm traffic. Walking routes are indicated on the information map in the parking area. To view wildfowl when the washes are flooded visitors are advised to walk behind the southern bank of the Counter Drain to a hide overlooking the Low Wash. A second hide is located in the wooded area near the junction of the farm track and River Glen bank. When the flooded washes are frozen, traditional fen skating is permitted by arrangement with the Lincolnshire Skating Association, and county and national events take place if the ice persists. For a location plan with adjoining reserves see page 187.

Description and Management

Baston Fen is a long tract of permanent pasture, which is flooded in winter and attracts large numbers of wildfowl. Near the entrance is an area of willow carr, which provides a range of nesting sites for birds. In summer a footpath leads through the carr into a wooded area. In winter the carr is flooded, and it is therefore necessary to follow the farm track and then turn left into the woodland to reach the hide. The carr once contained a hamlet of eight cottages and a windmill, but now its willows, osiers and alders provide additional shelter and nesting sites for birds, such as tits, finches and warblers.

The waymarked route continues along the river bank where the three washes can be observed. In the first of the three washes, the Low Wash, a high water-table has encouraged the growth of sedge beds. Adjoining the river bank are borrow-pits and sedge swamp, with stands of yellow flag, meadow-rue and meadowsweet. In the first two washes, meres have been excavated in the peat and attract summering wildfowl and waders. The edges of the meres have been colonised by typical fen plants, such as purple-loosestrife, water-plantain and marsh-marigold. The borrow-pits in the second and third washes contain such localised plants as greater bladderwort, water-violet, lesser water-plantain, greater spearwort and several species of stoneworts. Marsh orchids are found at the western end of the reserve. In total over 300 species of plants have been recorded in the reserve. Beyond the western end of the third wash, the bank overlooks a narrow belt of borrow-pits, marsh and scrub.

Birds are numerous and varied at most times of the year and over 160 species have been recorded. Of interest in the breeding season are shoveler, tufted duck, snipe, sedge warbler and, occasionally, garganey. In winter flocks of up to

1000 wildfowl, including teal, wigeon and mallard, and occasional Bewick's swan, pintail and goosander, can be seen, together with parties of snipe and redshank in March as the flood waters recede.

In summer there are various damselflies and a good variety of dragonflies, including hairy, migrant hawker, ruddy darter and four-spotted chaser. The reserve is also rich in other invertebrates, and over 70 species of non-marine molluscs have been recorded.

The Counter Drain Washes, of which Baston Fen is now the only part not cultivated, were used to hold excess flood water released from the River Glen and Thurlby Fen. The management of the reserve perpetuates winter flooding from early December to early March, and uncontrolled summer flooding sometimes occurs. The washes are cut for hay or grazed by cattle.

Baston Fen

Baston Fen

GR at E TF 145176

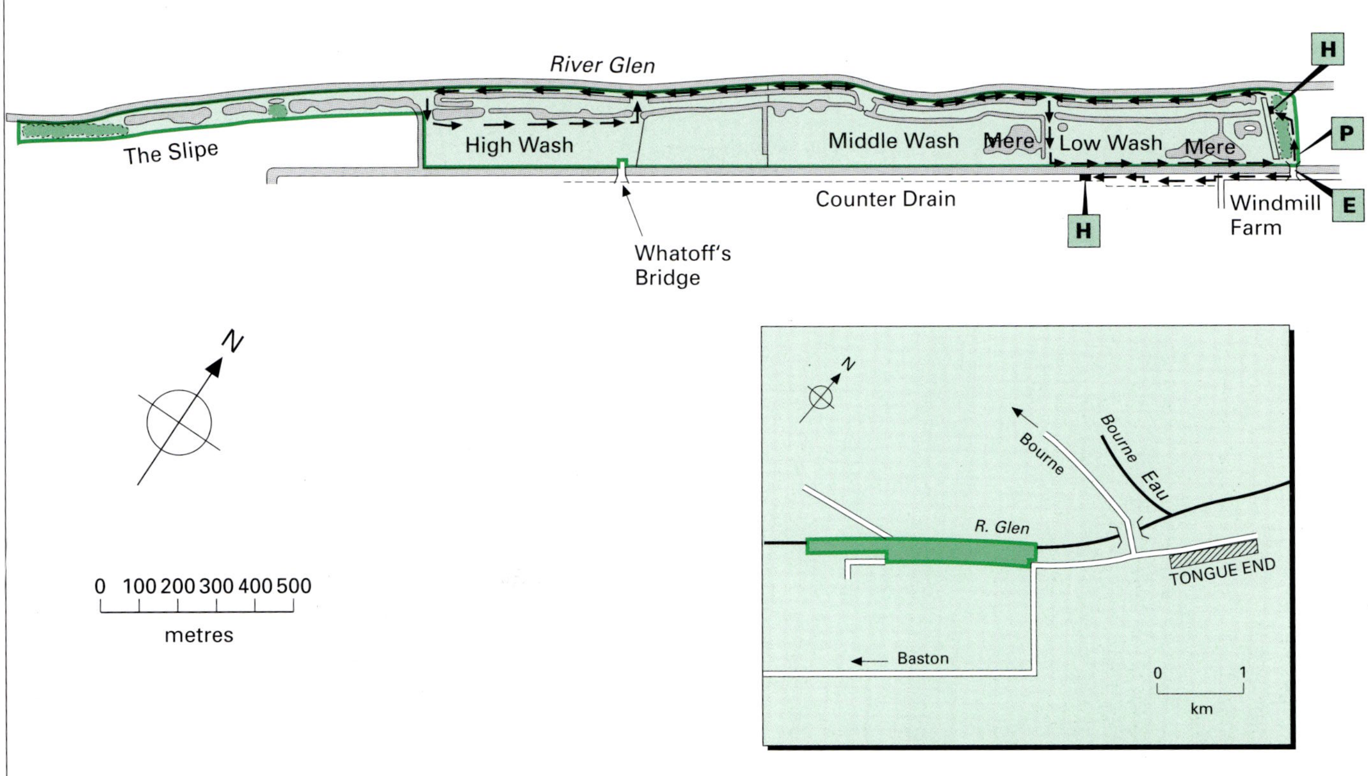

Boston Road Brick Pits

Gosberton
OS: 131 GR: TF 247319 1.8 hectares (4.5 acres)
Freehold 1967
Map ref: 7

Location and Access

The pits are located on the south side of the B1397 road at the east end of Gosberton village. Cars should not be parked where they will obstruct the drive to the adjacent house. Please also ensure that the gate is kept shut as sheep may be present in the reserve.

Description and Management

The reserve comprises part of an area of old clay pits with grassland, reedbeds, ponds and orchard. There is good aquatic habitat, with plants such as common reed, yellow water-lily and yellow iris. There is also a wide variety of nesting birds, including spotted flycatcher, reed and sedge warblers, coot and mallard. Finches, buntings and kingfisher are frequent visitors. Meadow brown, gatekeeper and ringlet butterflies may be seen in summer. In autumn red admirals feed on rotting fallen fruit, which also attracts fieldfares and redwings during the early winter months.

Management is primarily designed to maintain the intimate character of the old orchard with its mixture of fruit trees and adjoining broadleaved scrub. The sheltered grass paddocks are maintained by sheep grazing.

Apple trees, Boston Road Brick Pits

Boston Road Brick Pits

GR at E TF 247319

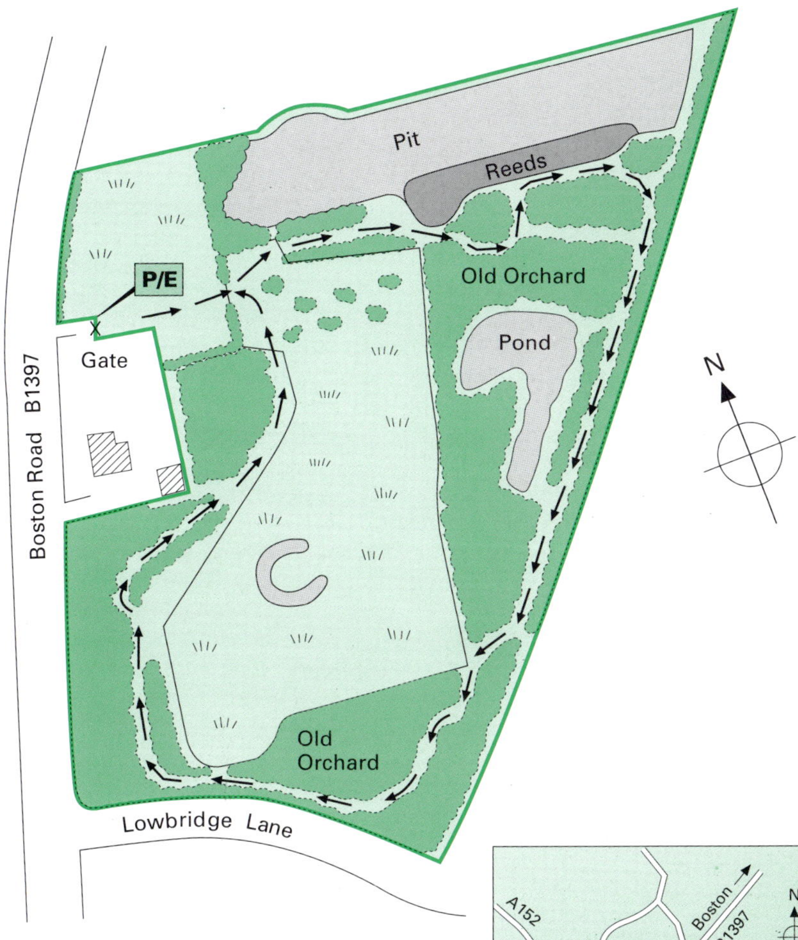

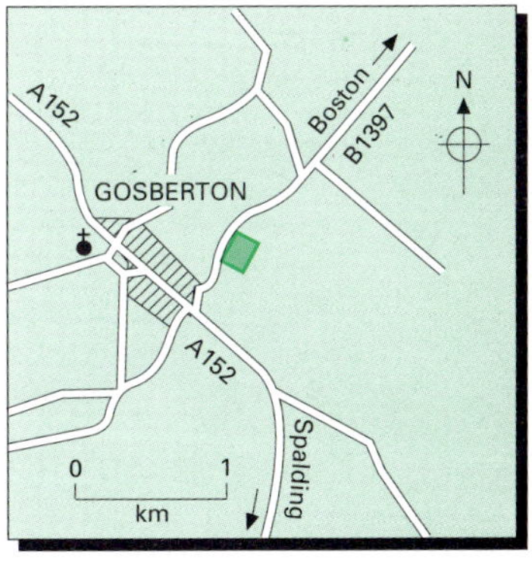

Boultham Mere

Lincoln
OS: 121 GR: SK 957713 19.1 hectares (47.3 acres)
Freehold 1988. Gift from British Rail Property Board
Map ref: 81

Location and Access

Access is from Tritton Way in Lincoln. Cars are best left at one of the car parks off Tritton Way. Follow the track alongside the drain, crossing the railway at a gated crossing. After about half a mile, there is a crossing point at a sluice over the drain that leads to the south-western bank of the reserve. Take care not to disturb wildfowl when climbing the bank.

Description and Management

Known formerly as Skewbridge Ballast Pit, the reserve lies within the City of Lincoln but, because it is bounded by the railway and Pyewipe Main Drain, it is a comparatively quiet and secluded sanctuary. As its old name suggests, it was excavated as a source of ballast for the construction of the railway. It has since matured into a most attractive lake with beds of reed and sedge, and emergent vegetation. There are mature trees and shrubs, mainly willow and sallow, flanking the railway. A hide overlooks the mere to the west of the reserve.

Covering over 47 acres, the reserve attracts many wintering wildfowl, including teal, shoveler, goosander and goldeneye. There are also many breeding birds, including up to 20 pairs of both reed and sedge warblers, two pairs of great crested grebes and three pairs of little grebes. Kingfisher and common tern also breed on the reserve. The reserve provides one of only two regular sites in Lincolnshire for wintering bittern. In recent years, scarce birds recorded include Savi's warbler, marsh warbler, purple heron, goshawk, Mediterranean gull, little gull, hobby and osprey. The reserve and adjoining drain are rich in dragonflies and damselflies with 12 species regularly recorded. Many butterflies may be seen in summer, including orange-tip, ringlet, large skipper, meadow brown and wall brown. Flowering plants of interest include lesser water-plantain and water-violet, both of which are 'red data species' featured in the Trust's publication *Nature in Lincolnshire: Towards a Biodiversity Strategy.*

Great crested grebe

Boultham Mere

GR at E SK 957713

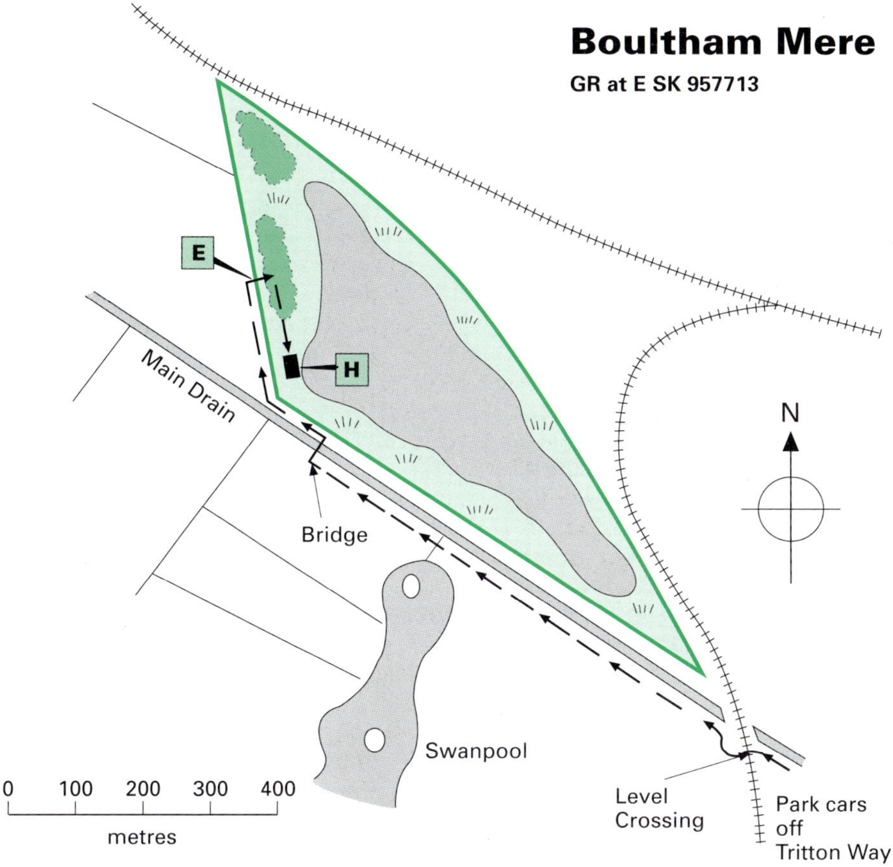

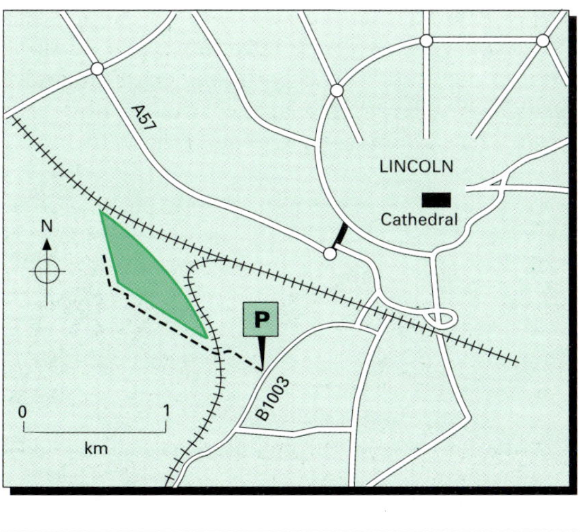

Burton Gravel Pits

Burton
OS: 121 GR: SK 943740 27.9 hectares (69 acres)
Leasehold 1971
Map ref: 8

Location and Access

The reserve is approximately 4 km (2.5 miles) north-west of Lincoln and lies on both sides of Fen Lane between Burton village and the A57. There is a parking area on the north side of the lane inside the main gate. This gate must be kept locked and the lock combination is made known by publication in the 'Lapwings' newsletter. Visitors are requested to keep to the waymarked route and avoid disturbance, particularly during the nesting season. Visitors should also be aware that some of the banks along the route are steeply sloping.

Description and Management

Although very near to Lincoln, this reserve is in a quiet, relatively secluded area. It comprises five pits with small areas of woodland and reedbed. The earliest excavations for gravel took place before 1920, and the latest about 1963. The marginal areas of shallows have developed a variety of swamp, marsh and carr vegetation. Rare or localised plants include greater spearwort, galingale, pillwort and flowering-rush. Nesting waterfowl include great crested grebe, little grebe and tufted duck. There are marsh and reedbed birds and, in the woodlands, green, great spotted and, occasionally, lesser spotted woodpeckers. A variety of ducks occur in winter. Rarer species of birds seen in recent years include red-necked grebe, red-crested pochard and black tern. Kingfishers occur throughout the year. Among bats identified are Daubenton's and noctule. Management includes scrub control, tree thinning, ride creation and path maintenance.

Waymarked Route 1.6 km (1 mile)

The route starts from the car park and encircles the main pit. Leave the car park on the westward path parallel to the lane for some 100 m. The broad path is bordered at first by bramble and rosebay willowherb. Look for butterflies, such as ringlet, gatekeeper and small heath. The route turns northwards along the wide ride flanked by mature beech and sycamore on the left (Kew's Holt) and wych elm, birch and ash on the right. Woodpeckers and five species of tits may be seen in this woodland area. Half way along the ride, a side path to the right provides a short-cut back to the car park along the side of the lake and through the birchwood, which attracts many songbirds, such as the blackcap. Continuing north beyond the ride the path narrows. On the right is a large pond with marginal bur-reed and willow coppice. A short distance past the pond, another side path to the right allows a return to the car park alongside the lake. There is a seat overlooking the lake at this point. The route continues northwards alongside the ditch on the right, rounds the north end of the lake and returns southwards along the Burton Main Drain, which harbours arrowhead, water dock, unbranched bur-reed and water-milfoil. There are also several species of damselflies and dragonflies to be seen. Wood small-reed grows on the banks of the drain. A path to the right leads to an observation hide, which overlooks the lake and provides the best place to view the waterfowl. This path leads back to the car park through a disturbed sandy area giving a good view of the islands. The path along the drain leads to a meadow area on the right. Cowslips and vetches can be seen in this grassland, while the fringing birches hold twayblade and umbellate hawkweed. Follow the mown path through the meadow back to the car park.

Candlesby Hill Quarry

Welton-le-Marsh
OS: 122 GR: TF 460682 1.7 hectares (4.3 acres)
Freehold 1964
Map ref: 9

SSSI

Location and Access

The quarry is situated on the east side of the A1028 Skegness–Louth road, about 1 km (0.6 miles) north of Gunby Corner roundabout. There is parking space for one car only in the entrance; otherwise park on the verge by the roadside. Visitors are advised to park well away from the dangerous bend. Cliff and scree should not be climbed and care must be taken not to trample vegetation unnecessarily in the summer. Beware of exposed lime kilns.

Description and Management

At the southern tip of the eastern edge of the chalk Wolds, this quarry was once the Gunby estate chalk pit. The lime kilns are probably older than those at Mill Hill Quarry in Claxby, and the quarry has been disused for longer. The chalk cliff exposure shows only the Lower Chalk. Several habitat types merge into one another. The track passes through tall hawthorn and ash wood, and scrub of dog rose and hawthorn, and finally emerges in an open grassland area, which is still being colonised. Most of the chalk flowers of the south-east Wolds (including four orchid species) are present, and woodland flora has developed in the ash wood. Breeding birds include garden warbler, whitethroat, spotted flycatcher and several species of finch. Butterflies are abundant and include common blue, small copper and wall brown.

Management consists of mowing and scrub control in order to maintain the various types of chalkland habitats and to diversify the woodland.

Dog rose

Candlesby Hill Quarry

GR at E TF 460682

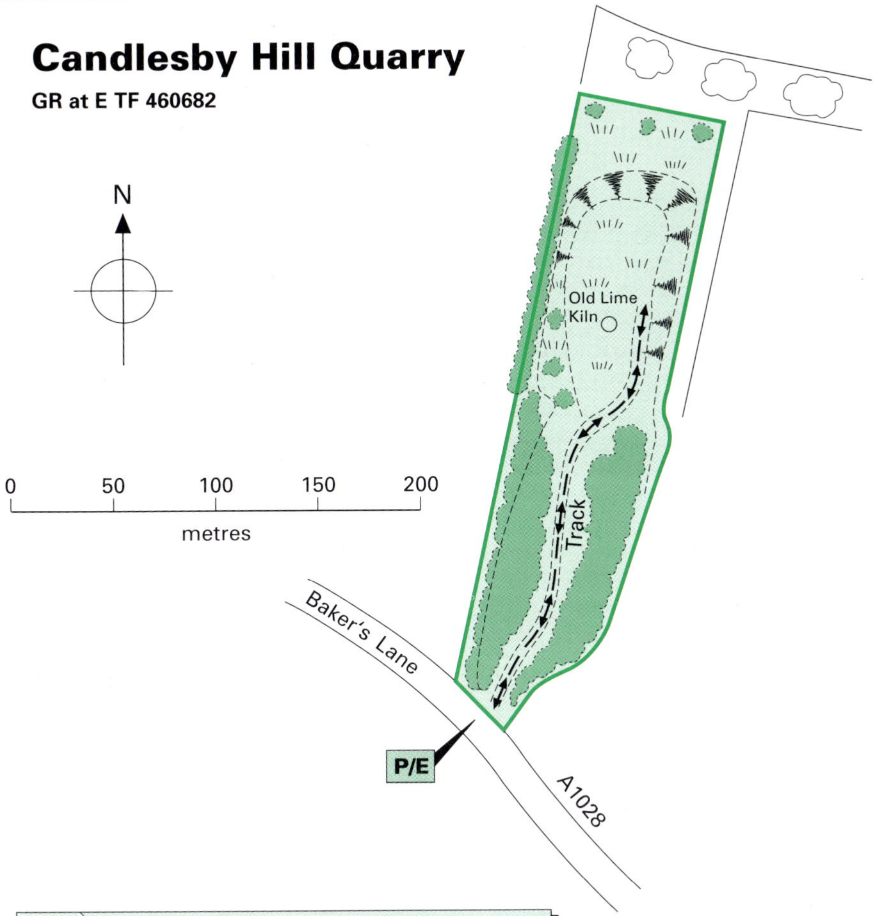

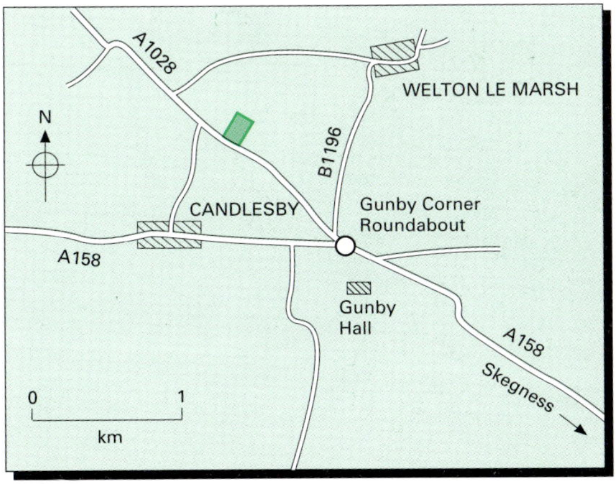

Chapel Pit

Chapel St Leonards
OS: 122 GR: TF 558740 3.2 hectares (7.8 acres)
Leasehold 1959
Map ref: 47

S SSSI

Location and Access

The pit is situated a short distance north of Chapel Point on the west side of the road. There is parking space on the opposite side of the road a little further north at Chapel Six Marshes. Visitors are requested not to walk round the pit. Views of the reserve may be had by walking down the short track from the road. An alternative viewing place may be reached by walking a few yards southwards.

Description and Management

Excavated for clay for the repair of the sea banks following the floods of 1953, the flooded pit has marginal reedbeds and aquatic plants, such as water-crowfoot and great reedmace. Fifteen species of ducks have been recorded, mainly in winter. Grebes, divers and wild swans are also often seen in winter. In summer breeding species include reed and sedge warblers, lesser whitethroat, and great crested and little grebes. In August and September thousands of migrating swallows and house martins roost in the reedbeds. Screens of willows round the banks of the pits have been planted in order to reduce disturbance to birds.

Swallows

Chapel Pit

GR at E TF 558740

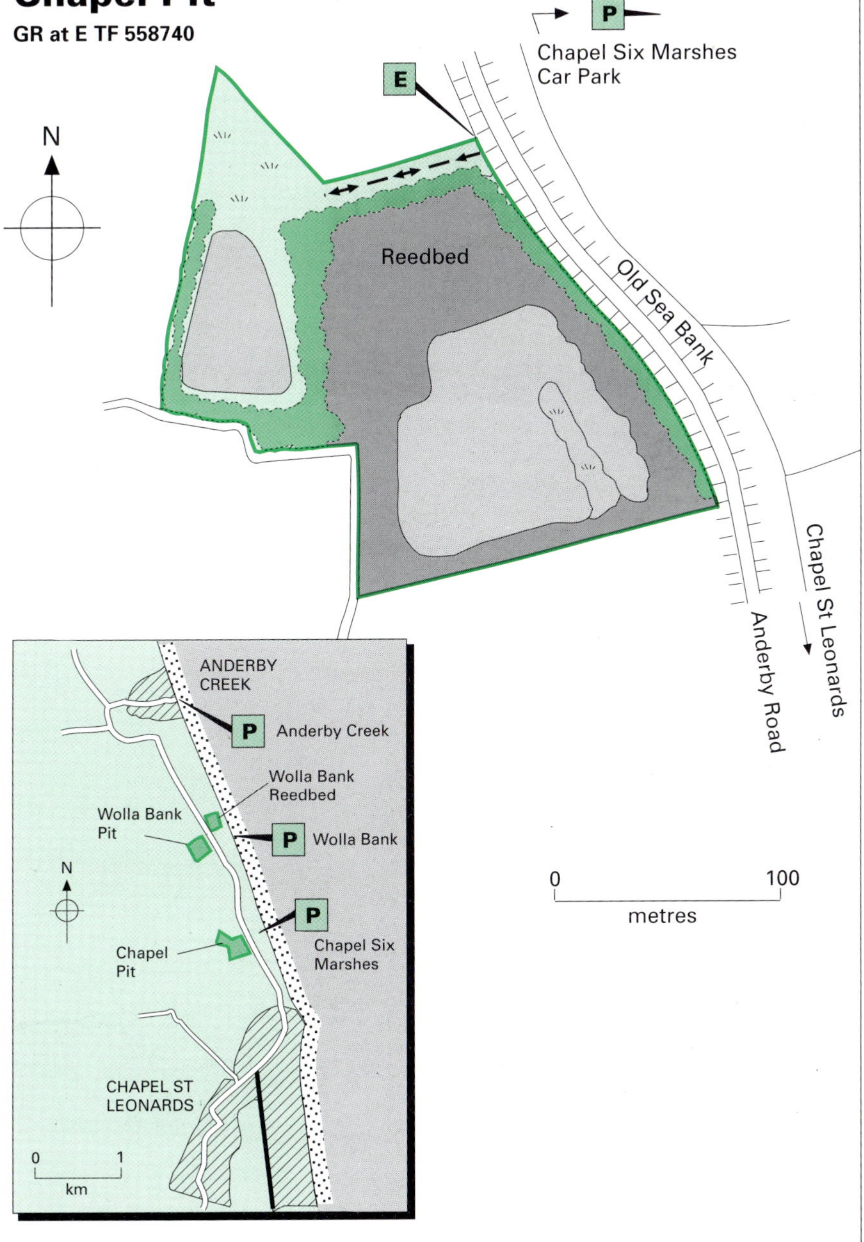

Clapgate Pits

Broughton
OS: 112 GR: SE 961109 1.0 hectares (2.6 acres)
Freehold 1996
Map ref: 104

Location and Access

The reserve is situated at the north-east corner of Broughton Far Wood on the B1208 about half a mile from its junction with the B1207 (Ermine Street). A public footpath skirts the western edge of the reserve. Access into the reserve will be provided either by a stile or a gate when fencing arrangements for grazing have been decided. Cars may be parked on the road verge.

Description and Management

Clapgate Pits consists of a long, disused quarry with a rich and varied limestone flora — the best in North Lincolnshire — which includes several plants rare in the county. Notable among these are pale St John's-wort, squinancywort and wall germander, a national Red Data Book species, often a garden escape but considered here to be of native origin. This was also formerly the most northerly site in Britain for pasqueflower but the plants were apparently dug up by vandals in 1969. In recent years scrub has enveloped much of the quarry; following a careful survey, this will be cleared and the area fenced for occasional sheep grazing in order to restore and maintain the grassland flora.

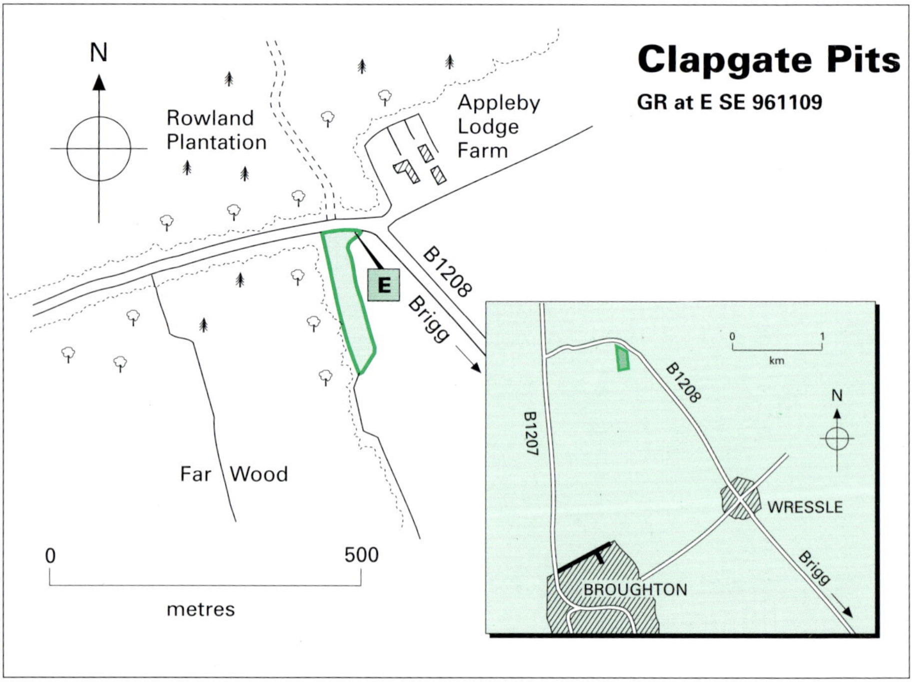

Claxby Spring

Claxby-by-Alford
OS: 122 GR: TF 450714 0.1 hectares (0.25 acres)
Management agreement 1978
Map ref: 10

Location and Access

This small site lies beside the road in the village of Claxby, near Willoughby, and is close to Mill Hill Quarry and Hoplands Wood reserves.

Description and Management

Burlands Beck rises from a chalk cliff and forms a shallow pool of clear, cold water held in place by a dam. Freshwater shrimps and caddis larvae abound, and the little flatworm *Crenobia alpina*, a rare species found in lowland Britain in the cool headwaters of streams, lives under the stones. There are many birds to be seen here, including grey wagtail and dipper in winter, and in summer spotted flycatcher in the backcloth of trees.

The Trust maintains the dam and clears out the pool as necessary.

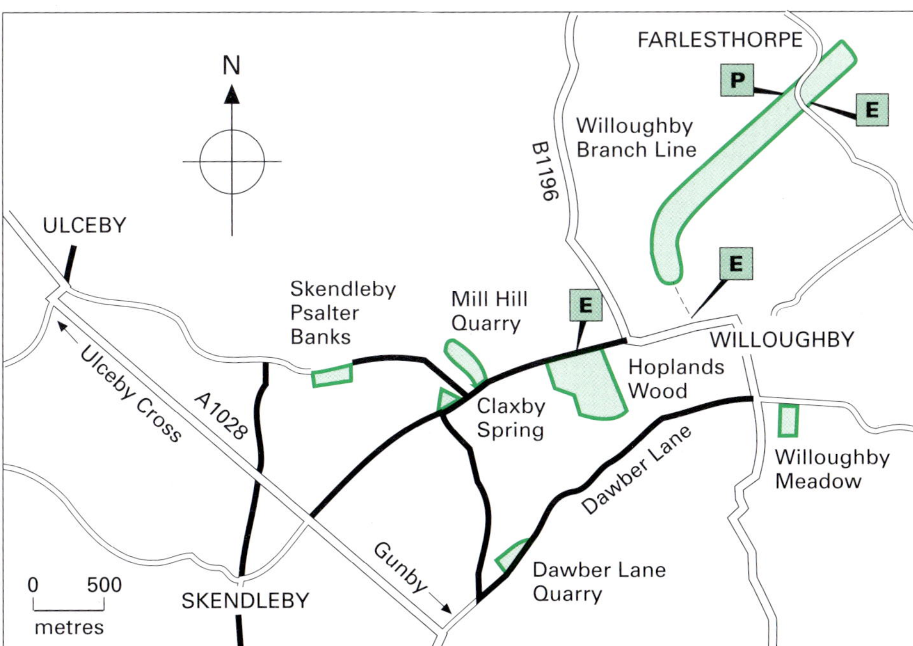

Covenham Reservoir

Covenham St Bartholomew
OS: 113 GR: TF 349955 (Grange Lane) 10.1 hectares (25 acres)
Management agreement 1974
Map ref: 11

The south-east corner of the reservoir is managed for wildlife conservation by agreement between the Trust, Anglian Water and the Covenham Water Sports Association.

Location and Access

The reservoir is situated to the south of the Marshchapel–Covenham road, approximately 8 km (5 miles) north of Louth. There is parking space at the north-west corner of the reservoir. There are two access points to the reserve: either on the north bank adjacent to the car park (TF 340962) up the steps and through the gate, or on the south bank from Grange Lane, following the bank approximately half way along, and up the steps and through the gate on the left opposite the small copse. Grange Lane is at the north end of Covenham St Bartholomew. Visitors may walk the whole perimeter of the reservoir; the perimeter track is now enclosed by a fence. Walking on or inside the wave wall is dangerous and therefore forbidden. The grass banks are grazed by sheep and therefore dogs should be kept on leads. Please observe the safety notices in the car park and at various points on the perimeter track. The Trust has erected a hide on the south-east side for the use of members; please take care when entering and leaving and make sure that the doors are properly secured.

Description and Management

The reservoir is artificially embanked for its entire length. The sides are steeply shelving and there is no marginal vegetation. Nevertheless, it attracts large numbers of wintering wildfowl and other aquatic and waterside birds. It is used extensively as a daytime roost by surface-feeding ducks, but its main attraction is to diving ducks. In winter many tufted duck, pochard and goldeneye occur, with smaller numbers of sea duck. Red-throated divers are occasional winter visitors, and four species of grebe have been seen on a single day. In addition to the water birds, migratory waders and passerines, including rarities, feed at the water's edge. Yellow wagtails are numerous from mid-April to mid-May. On the north bank spotted orchids, cowslips and common centaury can be found. The plantations on the north and south banks and the open ditches running along the bottom of the east and west banks add to the diversity of habitat.

Covenham Reservoir

GR at E TF 349955 (Grange Lane access)

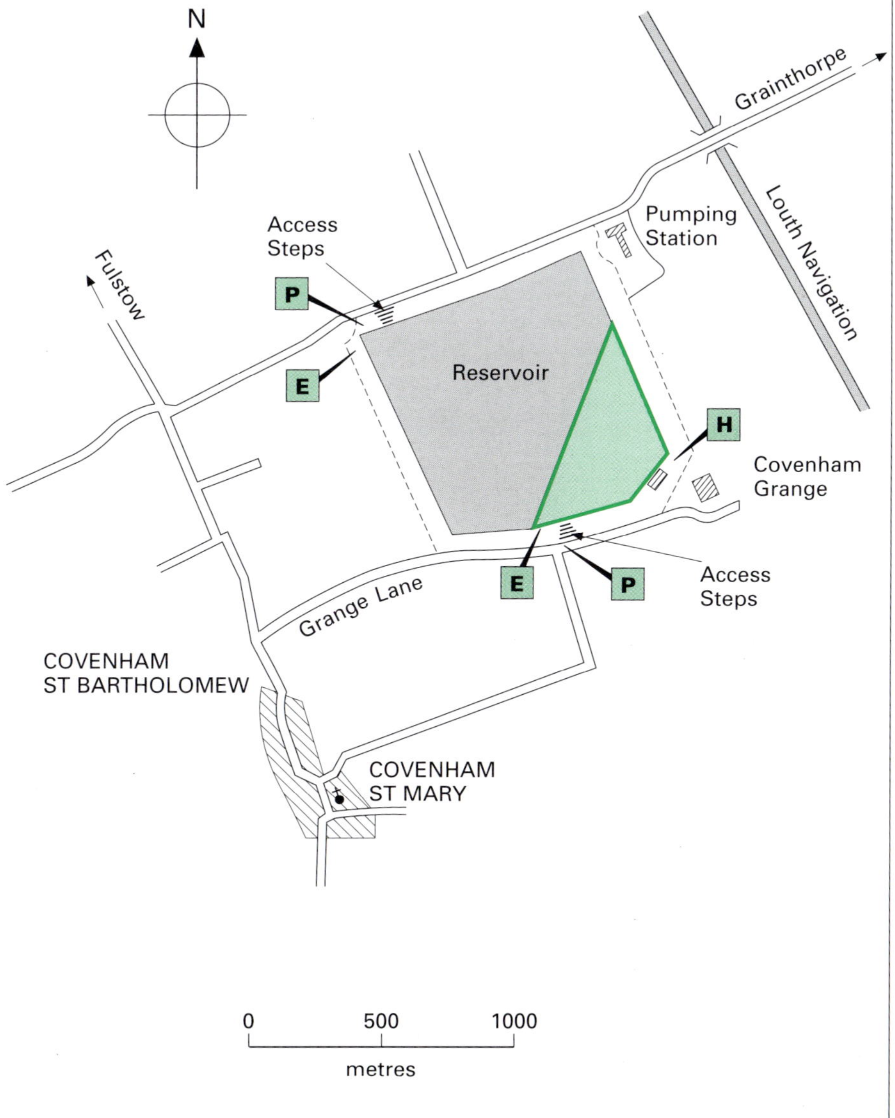

Crowland Ponds

Crowland
OS: 131 GR: TF 229107 to 232125 2.4 hectares (5.9 acres)
Leasehold 1992
Map ref: 98

Location and Access

The ponds lie 1.6 km (1 mile) north-east of Crowland at the bottom of the east bank of the River Welland. An unclassified road runs along this bank and all the ponds may be viewed from this road.

Description and Management

There are 35 ponds in total. These narrow ponds, forming a long line, were once borrow-pits, which then became MAFF experimental fish breeding ponds. Coot, moorhen and mute swan usually breed in the ponds. Reed and sedge warblers also nest here in good numbers. The ponds make an attractive and useful feature in an otherwise rather featureless landscape.

Mute swan

Crowland Ponds

GR at TF 229107–232125

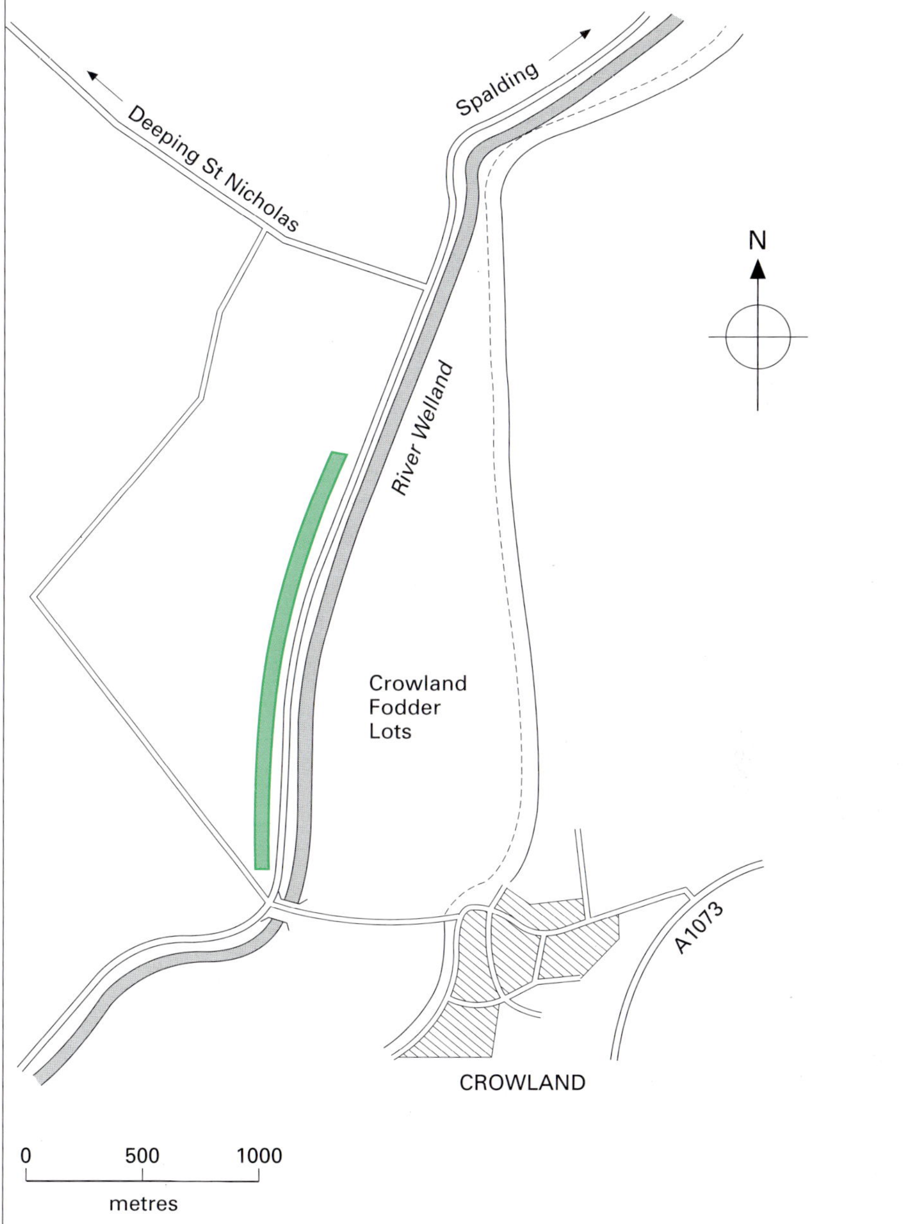

Crowle Moor

Crowle
OS: 112 GR: SE 759145 & SE 756137 (car parks) 162.6 hectares (401.7 acres)
Freehold 1971
Map ref: 12

SSSI

Location and Access

Crowle Moor lies in the north-west corner of the Isle of Axholme and is reached from an unclassified road running north-west from Crowle village. It is part of the Humberhead Peatlands National Nature Reserve. For the southern area, park at grid reference SE 756137 where there is a visitors' car park from where a waymarked route of about 2 km (1.25 miles) starts. For the northern area, where there is a longer waymarked route of some 6 km (3.75 miles), park at the roadside at SE 759145. It is advisable to follow the waymarked routes, as the reserve is large and complex. There is a particular danger of fire and therefore visitors are requested not to smoke while on the reserve.

Description and Management

The still extensive tract of Hatfield, Thorne, Goole and Crowle Moors is but a remnant of the vast complex of moor, bog and fen that once surrounded the head of the Humber estuary and included much of Lincolnshire's Isle of Axholme. Drainage and clearance for agriculture from the 17th century onwards left the present moors as 'islands' of raised peat bog, formerly used as turbaries by the villages around. It is likely that the peat on Crowle Moor was not as thick as that on the Yorkshire moors (though it still averages about 3 m) and certainly less seems to have been extracted. The Moor is divided into 'ribbons' running into the moor from the warpings — the cultivated land near Crowle village — and represents holdings carved out of once common turbary.

The reserve is one of the richest lowland peat vegetation areas in the north of England. The higher, drier areas carry heather, bracken and birch scrub; the wetter parts have reedbeds, cottongrass, *Sphagnum* bog, willow carr and open water. Rarer plants include bog rosemary, dune helleborine and greater yellow rattle.

These varied habitats support a rich bird, mammal and insect fauna. The large heath butterfly occurs here at the south-eastern limit of its range in Britain. More than 30 breeding birds have been recorded, including long-eared owl, woodcock, nightjar and tree pipit. Grass snake and adder are present.

Management is chiefly concerned with the maintenance of a high water-table and the control of scrub encroachment. Extensive areas of wet heath have been cleared of invading birch scrub and are now grazed by the Trust's flock of Hebridean sheep.

Crowle Moor

GR at SE 759145 & 756137 (car parks)

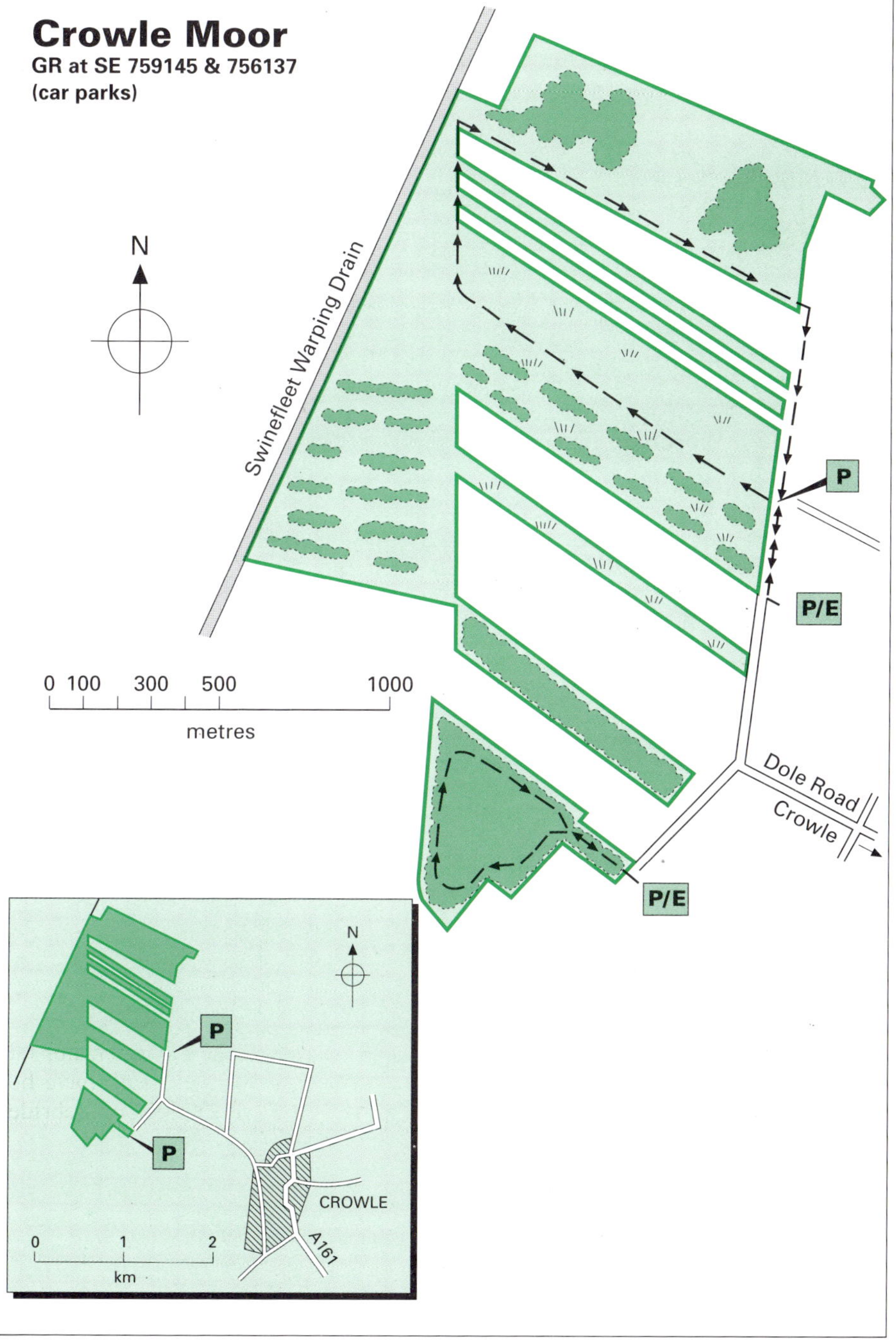

Dawber Lane Quarry

Welton-le-Marsh
OS: 122 GR: TF 454704 0.24 hectares (0.6 acres)
Management agreement 1978
Map ref: 13

Location and Access

The quarry is situated on the north side of Dawber Lane, which links the A1028 with the B1196 near Willoughby. As the site is very small and difficult to enter, access is by special permit only. For a location plan with adjoining reserves see page 51.

Description and Management

A small chalk quarry with a varied and interesting chalk flora. After initial clearance of dumped rubbish, the chalk grassland vegetation has been managed by occasional mowing and by checking the advance of scrub. The reserve adjoins the Dawber Lane protected roadside verge.

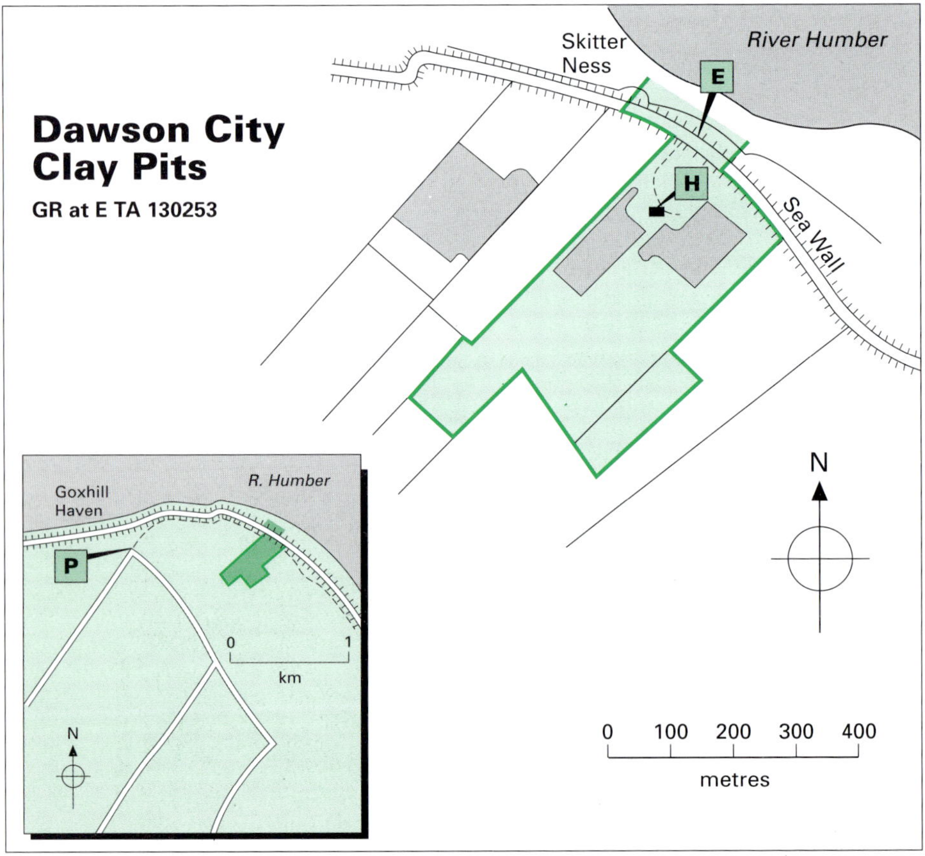

Dawson City Clay Pits

Goxhill
OS: 113 GR: TA 130253 16 hectares (39.6 acres)
Freehold 1986. Part gift of Mr JFD Hargreaves
Map ref: 76

SSSI

Location and Access

The reserve is situated north-east of Goxhill adjoining the River Humber. Cars may be parked at Goxhill Haven. Walk along the Humber Bank eastwards for about 1 km (0.6 miles). The reserve is the second pit on the right and extends to the foreshore at Skitter Ness, which is part of the Humber Bank SSSI. The entrance is by the stile at the north-west corner. A hide has been constructed for visitors' use and is located along the waymarked route between the two pits. Visitors are requested to proceed quietly in order not to disturb the birds and other wildlife. For a location and reserve plan see map opposite.

Description and Management

The reserve consists of two disused clay pits with associated reedbeds and adjoining scrub and pasture. The pits are particularly important for a wide range of birds, not only as a breeding site but also as a migration staging post. There are 108 species of birds recorded on the reserve with over 20 species breeding, including reed and sedge warblers, water rail, garganey and, in the past, bittern. Snipe are regular visitors. Migrant waders include green and wood sandpipers, redshank, greenshank and ruff. Numbers of surface-feeding ducks, such as shoveler, teal and mallard, are at their greatest in late winter and early spring. Vagrants have included spoonbill, green-winged teal, Bewick's swan, little bittern, red-crested pochard and pectoral sandpiper. A 9-hectare (23-acre) block of old pasture was added to the reserve in 1991. This is a good area for curlew, the numbers of which in winter are often in excess of 200, and for other waders, especially when the fields are flooded in winter. The reserve extends to part of the foreshore, from where both passage and wintering waders may be seen.

Management consists of grazing and occasional hay-cutting of the pasture. Fencing has been erected and trees and shrubs planted to screen the hide.

Water rail

Digby Corner

Digby
OS: 121 GR: TF 088551 0.36 hectares (0.89 acres)
Management agreement 1992
Map ref: 94

Location and Access

This small triangular piece of railway land is bounded by the Lincoln–Sleaford line, and is situated about 0.4 km (0.25 miles) east of Digby on the road to Walcott. The access gate is a short distance beyond the railway bridge, on the right side of the road.

Description and Management

Digby Corner is a little oasis in an arable farming area. It is situated on the line of the old road that was diverted when the railway bridge was constructed. The soils are derived from the limestone on the most easterly projection of the Lincoln Edge. The calcareous soils support an interesting variety of wildflowers, including herb-Robert, cowslip, bird's-foot-trefoil, bladder campion, salad burnet, rue-leaved saxifrage, field madder and lady's bedstraw. Although mainly a grassy area, there are dense areas of scrub of hawthorn, privet, purging buckthorn, dog rose and bramble. In autumn plentiful berries and other fruits attract many birds. Breeding birds include lesser whitethroat, blackcap, linnet and yellowhammer. The reserve is also good for butterflies, such as common blue, small skipper, orange tip, brimstone and ringlet.

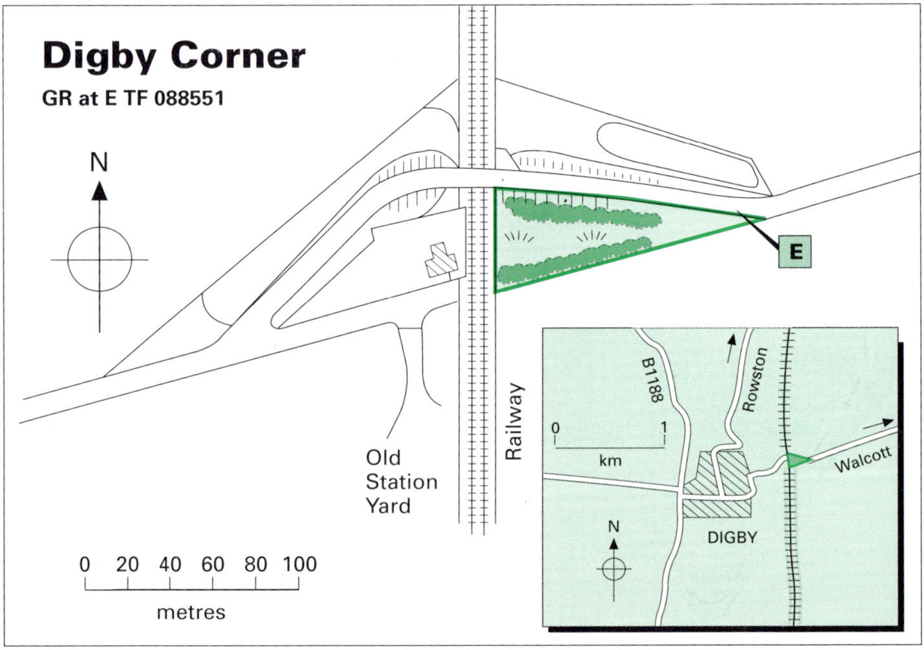

Dole Wood

Thurlby
OS: 130 GR: TF 094161 2.7 hectares (6.6 acres)
Leasehold 1975
Map ref: 14

SSSI

Location and Access

The wood lies to the south-west of the village of Thurlby, approximately 4 km (2.5 miles) south of Bourne. Cars may be parked on the roadside along Obthorpe Lane, which runs south from Thurlby and roughly parallel to and west of the A15. The track to the reserve lies to the west of Obthorpe Lane, about 500 m from the village. It is a further 500 m walk down the track to the entrance. Visitors are requested to keep to the perimeter pathway.

Description and Management

Dole Wood is a fragment of the formerly extensive primary woodland cover of South Kesteven. The wood consists mainly of oak standards with hazel coppice. There are also ash, field maple, wych elm and wild service tree. Both common and midland hawthorns can be seen in the understorey. Bluebell, wood anemone and stitchwort are the most abundant species in the ground flora. Wood-sorrel, wood sage, false brome, enchanter's-nightshade, bugle, yellow archangel and sweet woodruff can also be seen. Birds include lesser and great spotted woodpeckers, treecreeper, whitethroat, blackcap, willow warbler and spotted flycatcher. Over 120 species of moth have been identified within the reserve.

Management consists of bramble control, re-establishment of a coppice rotation and maintenance of some permanent glades.

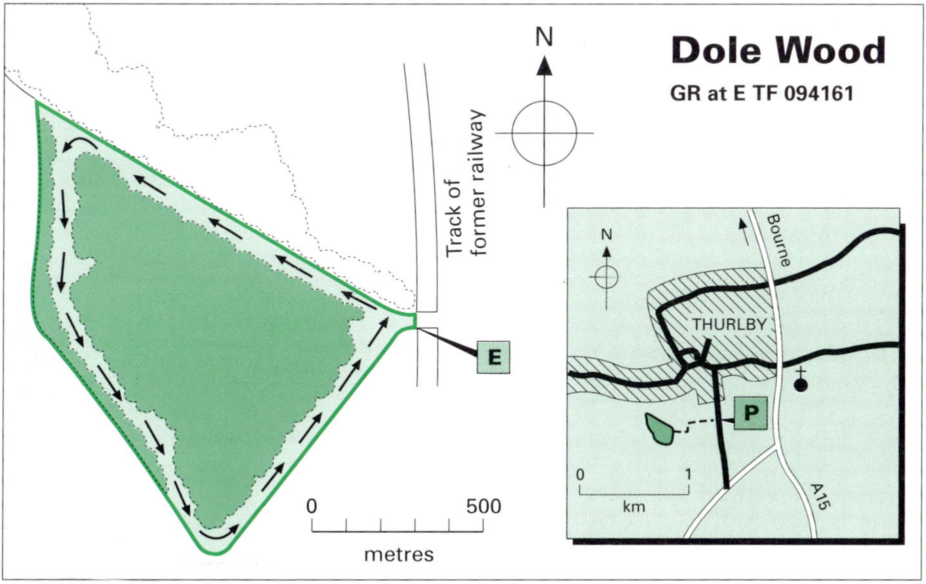

Donna Nook

North Somercotes and Skidbrooke-with-Saltfleet Haven
OS: 113 GR: TF 422998 (Stonebridge entrance) 1149.7 hectares (2841 acres)
Part leasehold, part management agreement 1978
Map ref: 15

Location and Access

The reserve covers more than 10 km (6.25 miles) of coastline between Grainthorpe Haven in the north and Saltfleet in the south where it borders the Saltfleetby–Theddlethorpe Reserve. There are several access points off the main A1031 coastal road. There are parking facilities at (from north to south): Stonebridge (GR: TF 422998), Howden's Pullover (GR: TF 449952), and Sea Lane, Saltfleet (GR: TF 456944). Public access is also possible at Merrikin's Pullover (GR: TF 445958) and Saltfleet Haven (GR: TF 467935), but these are unmetalled roads without parking facilities. Visitors should note that the Pyes Hall, Warren Farm, Red Farm and Coastguard Cottage entrances are all private and should not be used.

Visitors should be made aware that the Ministry of Defence still maintains part of the area as a bombing target range and UNDER NO CIRCUMSTANCES SHOULD ANYONE ENTER THE BOMBING AREA WHEN RED FLAGS ARE FLYING. However, most of the dune area is accessible at all times.

Description and Management

The reserve consists of dunes, slacks and inter-tidal areas. Coastal processes, particularly sand and mud accretion, alter the natural features from year to year, and sand from the beach and offshore sandbanks is blown inland by easterly winds to form dune ridges. Deposition of material from the River Humber has resulted in mudflats and saltings. The advancing dunes have trapped areas of saltmarsh behind them, and these areas have gradually become less saline, allowing an interesting plant community to develop. On the landward side, the reserve is bounded by a sea bank erected after the floods of 1953. The reserve is rich in bird life: 47 species of bird breed regularly and the area is famous for more uncommon passage migrants and rarities; over 250 species have been recorded in total.

The dune system is dominated by marram and sand couch, which play important stabilising roles, particularly in the early stages of dune development. Flowering plants here include yellow-wort, and bee and pyramidal orchids. Sea-buckthorn is encroaching and forms dense clumps in some areas. Among mammals, fox, badger, stoat and weasel are all present, and three species of shrew have been identified. Common lizards also inhabit this part of the reserve. In summer breeding birds of the dunes include red-legged partridge, dunnock, whitethroat, linnet, skylark, yellowhammer and tree sparrow, while in winter the sea-buckthorn berries attract large numbers of fieldfare, redwing and starling. Woodcock, hen harrier and short-eared owl are also regular winter visitors. Within the dunes there are attractive dune-slack meadows where many wildflowers, including marsh-orchids, can be seen.

Between the advancing dunes and the sea bank there are areas of saltmarsh and open lagoons originally formed by excavation work for the sea bank. Although

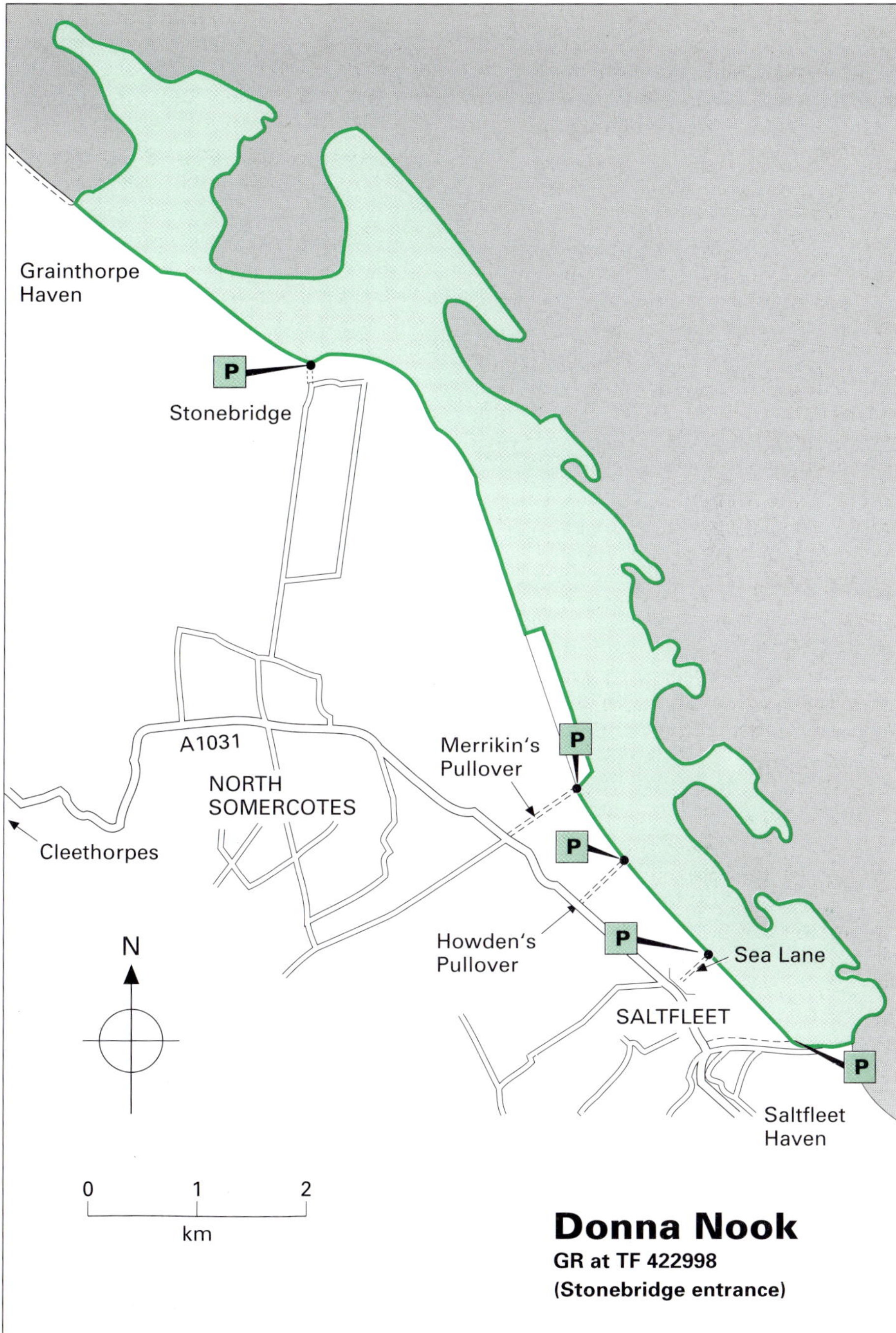
Grainthorpe Haven
P
Stonebridge
A1031
NORTH SOMERCOTES
Cleethorpes
Merrikin's Pullover
P
P
Howden's Pullover
P
Sea Lane
SALTFLEET
P
Saltfleet Haven
N
0
1
2
km
Donna Nook
GR at TF 422998
(Stonebridge entrance)

dominated in places by sedges and rushes, the open water with its muddy margins attracts a wide variety of migrating waders and ducks. The lagoons also provide a breeding area for small numbers of coot, little grebe and moorhen and a substantial population of reed bunting and meadow pipit. On the seaward side of the dunes there is a small area of saltmarsh.

The saltmarsh quickly gives way to inter-tidal mudflats on which grow the locally renowned samphire beds. The mudflats attract a wealth of birds: substantial numbers of brent geese, shelduck, twite, lapland bunting, shore lark and linnet gather in winter, together with large flocks of knot and dunlin accompanied by a wide variety of other waders in smaller numbers.

By far the largest part of the reserve consists of sandflats, and the raised sand bars are of special interest as they not only provide a hauling-out point for grey and common seals, but also support one of the few breeding colonies of little tern in Lincolnshire. Ringed plover and oystercatcher also nest in this area. Between autumn and spring huge flocks of gulls gather, while large congregations of terns, particularly sandwich tern, are a feature of late summer.

The reserve has one of the largest and most accessible breeding colonies of grey seals in the UK. From October to December the Trust mounts a wardening service to protect the seals and to organise watching facilities for visitors.

Management has concentrated on managing dune grassland, a substantial area of which is mown to create conditions suitable for a more diverse flora and fauna. Surveys are conducted annually to monitor changes in the flora and fauna. Nesting shore birds are monitored in the spring and early summer, and breeding grey seals are monitored in autumn and early winter. In 1988 an extensive shallow lagoon was created north of Merrikin's Pullover. Further south, work is in preparation under the Reserves Enhancement Scheme to improve an area of grassland and saline pools.

Grey seals

Duke's Covert & Copper Hill

Wilsford
OS: 130 GR: SK 980422 1.8 hectares (4.4 acres)
Freehold 1986
Map ref: 71

SSSI

Location and Access

Duke's Covert lies in the triangle of land at Copper Hill between Ermine Street and the road to Heydour, about 1 km (0.6 miles) to the south of Ancaster. The reserve entrance is some 200 m along the road to Heydour from the Ermine Street junction. There is parking space at the roadside. For a location plan with adjoining reserves see page 25.

Description and Management

The roadside verges at Copper Hill are among those protected under the scheme operated by the Trust and Lincolnshire County Council (PRV 024). They have a particularly rich limestone flora with much rock-rose, horseshoe and kidney vetches, spiny restharrow, purple milk-vetch and fairy flax. Butterflies include brimstone and common blue. This was the last known Lincolnshire locality for the chalkhill blue butterfly, and is the most northerly point for man orchid in Britain.

The adjoining Duke's Covert is old limestone grassland, which accommodates landing lights for the nearby Barkston Heath airfield. In recent years it had become overgrown and invaded by scrub of gorse and hawthorn, but with careful removal of scrub, and with mowing and grazing, the interest is being restored. Many limestone flowers thrive here, including common and greater knapweed, mignonette, pepper-saxifrage, dropwort and columbine, a scarce plant in Lincolnshire. In spring there is a colourful display of cowslip, hairy violet and some early-purple orchid. As spring rolls into summer the flora includes slender St John's-wort, rock-rose, horseshoe vetch and small patches of purple milk-vetch and lesser meadow-rue.

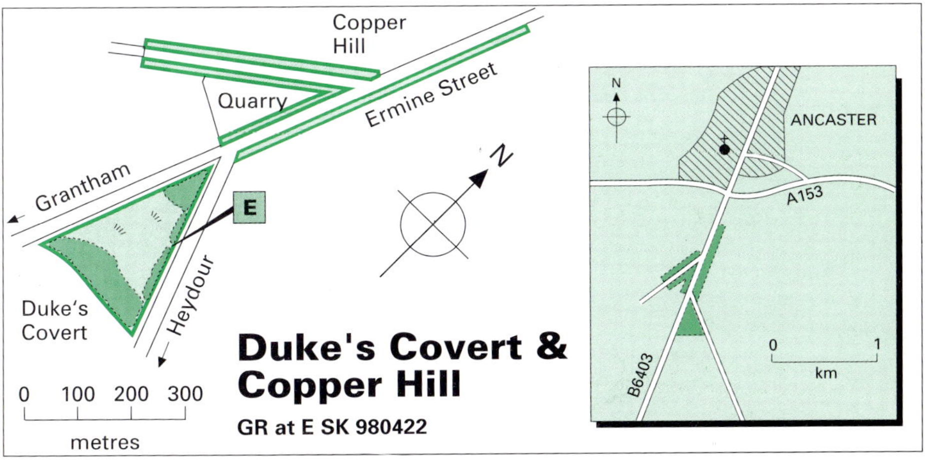

Eastfield Road Railway Embankment

North Killingholme
OS: 113 GR: TA 152171 0.7 hectares (1.7 acres)
Management agreement 1979
Map ref: 16

S

Location and Access

This small site is situated on the railway embankment of Eastfield Road, North Killingholme. Due to the sensitivity of the site and the terms of the Trust's agreement to manage railway land, access is by special permit only.

Description and Management

The grassland of the reserve represents a type of habitat with specialised plant species that has all but disappeared through changing land use. Plants of interest include bee orchid, yellow-wort and restharrow, and over 50 species have been recorded.

Management consists of controlling scrub and annual autumn mowing.

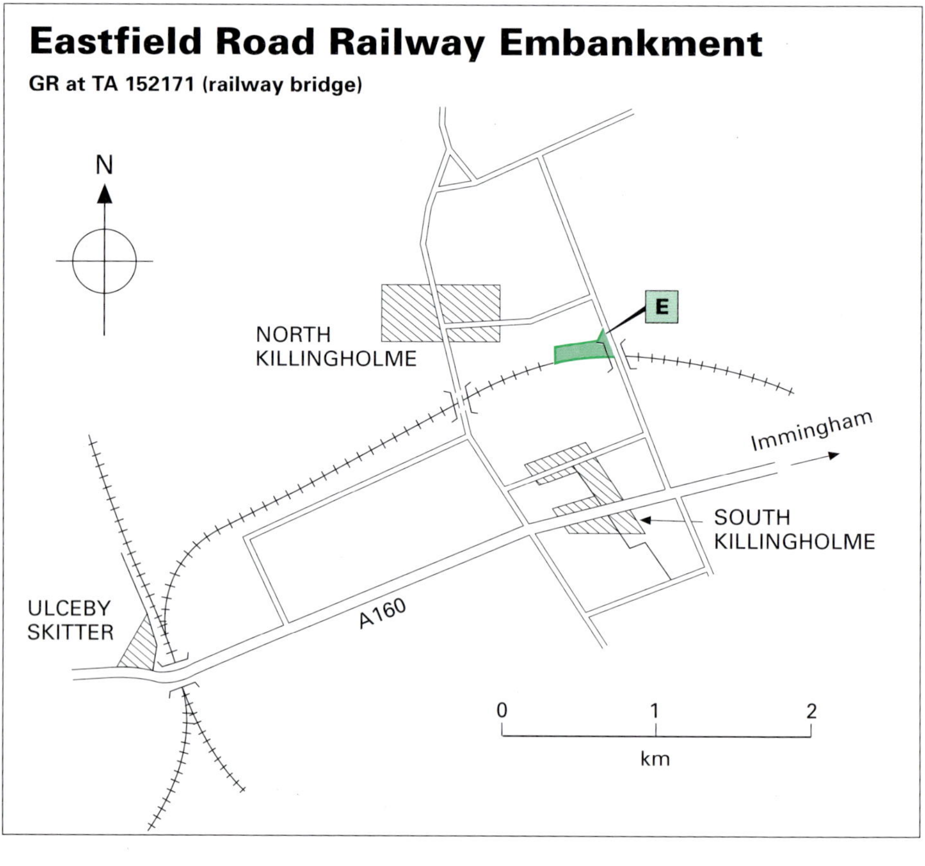

Epworth Turbary

Epworth
OS: 112 GR: SE 758036 33.2 hectares (82 acres)
Freehold 1958
Map ref: 17

SSSI

Location and Access

The Turbary is on the Wroot road, about 2.5 km (1.5 miles) west of Epworth. The entrance is near the bridge over Skyer's Drain. Parking is available through the gate, which should be kept closed, or on the verge adjoining the reserve; it is advisable to park well away from the corner. Visitors are requested to keep to the waymarked paths and to use the hides when viewing the open area. In order to avoid disturbing birds on the ponds, please do not climb on the banks.

Description and Management

One of the few relicts of raised bog in the whole of the Trust's area, this reserve is of considerable historic and scientific interest. The area has been extensively dug for peat in the past but areas of active *Sphagnum* bog still exist, with common cottongrass and cross-leaved heath. There are also areas of reed swamp and mixed fen vegetation, where the rare fen sedge, *Cladium mariscus*, which was once widely used for ridge thatching, can be seen. Other parts of the reserve have a series of fen and wet heath conditions and a considerable area of birch woodland of varying age. Plants in the fen areas include yellow and purple-loosestrife, meadow-rue, sneezewort and devil's-bit scabious. Among moths the wood tiger is well established. Breeding birds include tree pipit, warblers, finches, green and great spotted woodpeckers and woodcock.

The principal management problems at the reserve arise from the lowering of the water-table due to drainage of surrounding land and the consequent encroachment of birch scrub onto the wet heath and fen areas. Removal of birch has been the main management activity, together with the creation of ponds and small lakes, which are linked by channels, in the wetter areas. This work has resulted in rapid colonisation by *Sphagnum* mosses on bare sand below water-level. The ponds have attracted eleven species of breeding dragonflies and damselflies. Various bird species new to the reserve have also been attracted to these areas, including greenshank, green sandpiper and little grebe. Two areas that have been cleared of scrub are now enclosed by sheep fencing, and during the summer months (May to September) these areas are grazed by the Trust's flock of Hebridean sheep. This rare and hardy breed is providing an effective means of controlling birch re-growth.

Waymarked Route 2 km (1.2 miles)

From the car park follow the narrow path in a westerly direction and observe the many flowers along it. This corner of the reserve is one of the most sheltered and on fine days there is often an abundance of butterflies and other insects. The path continues, following the southern side of the reserve, for about 700 m. A variety of trees flank the path, including alder buckthorn, rowan, oak, elder, sallow, aspen, hawthorn and a great deal of birch. Plants of special interest include creeping willow and fen sedge.

About half way along this side, a path leads off to the right; keep to the right and after 40 m you can view the open area from Steve's Hide, which overlooks Steve's Pond, created as a memorial to Steve Clarke, a young Trust member from the Isle, who was tragically killed in a road accident. The area attracts a variety of birds: hobby, marsh harrier, greenshank, green and common sandpipers are among the species recorded. More regular sightings include teal, little grebe, sedge warbler, tree pipit and sparrowhawk.

On returning to the waymarked route you will reach the south-west corner of the reserve where the path turns right and leads north. Great spotted woodpeckers have bred in this area on several occasions. The surface becomes raised and the sandy soil contains old fox earths. The path then runs parallel to another deep drain which joins Skyer's Drain at the south-west corner. Both these drains have been considerably deepened during the last 30 years, and this has contributed significantly to the drying out of the reserve. As you follow the path look out for birds including willow tit, long-tailed tit, reed bunting and willow warbler. Occasionally corn buntings can be heard singing on the adjoining farmland. In the autumn and winter large flocks of rooks, crows and jackdaws assemble in the fields at dusk before flying into the reserve to roost.

At the north-west corner the path turns right and leads eastwards. After about 150 m the path crosses a small ride, which has been cut to stop trees touching the overhead power lines. The path continues eastwards through birch woodland with small clearings, where common reed still persists and cross-leaved heath can also be seen. Sometimes in winter long-eared owls can be observed roosting close to the path.

As you approach the eastern end of the reserve, the path turns right and heads south, leading to the second hide overlooking Pantry's Pond. As you look out of the hide to the right the ground rises slightly, and ling and cross-leaved heath have been encouraged by baring the ground. Looking to the left the ground is lower, and cottongrass and fen-sedge flank the edge of the pond. In winter you may be lucky enough to see a hen harrier. Other birds include yellowhammer, linnet, jay and magpie.

There is a loop path off the waymarked route that goes through an area of open heathland to the east of Pantry's Pond. The heathland area is now fenced but a stile at each end of the loop enables access to the path. If sheep are present members may still use the path but caution is needed so as not to disturb them. The path is accessed by turning left about 100 m before the hide at Pantry's Pond (see map). This path rejoins the waymarked route just after Pantry's Pond. Turn left and after 60 m turn left again onto the narrow path, which leads back to the car park.

Epworth Turbary

GR at E SE 758036

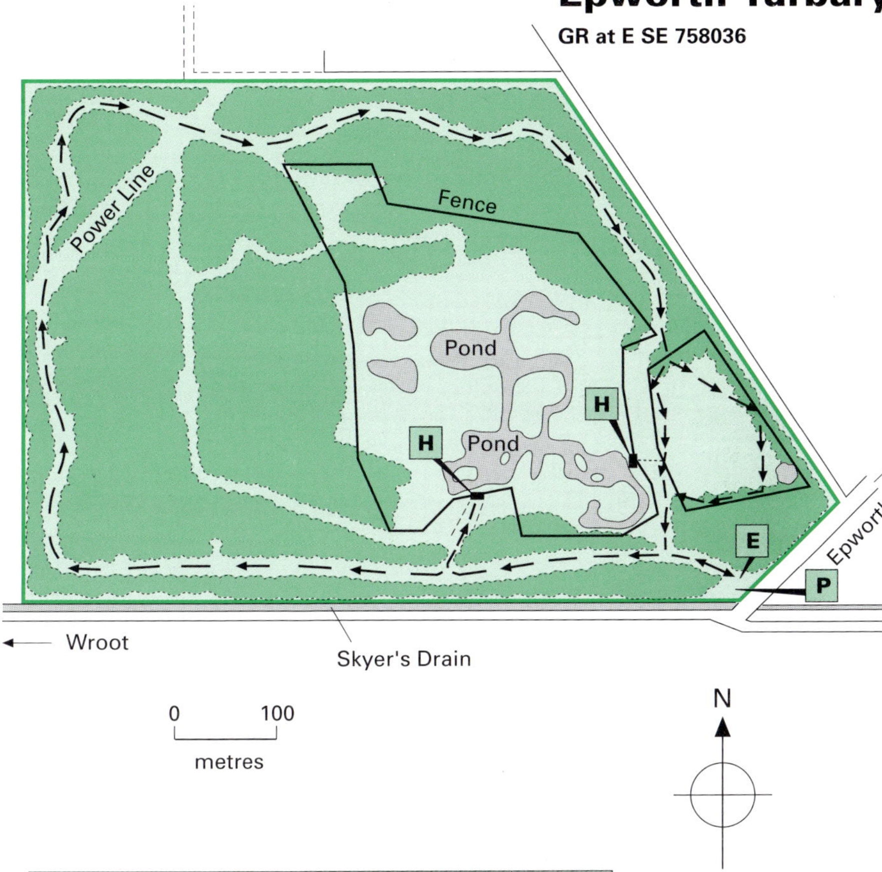

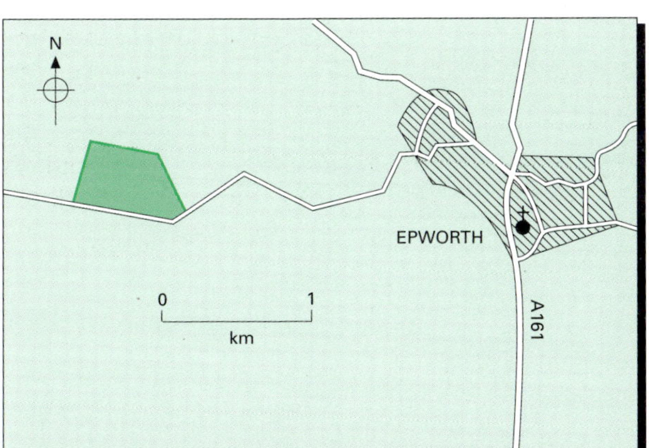

Ewerby Pond

Ewerby and Evedon
OS: 130   GR: TF 145499   0.9 hectares (2.3 acres)
Leasehold 1975
Map ref: 18

Location and Access

The reserve is situated on Ewerby Waithe Common on the lane from Halfpenny Hatch (near South Kyme) to Ewerby Thorpe. The lane separates the pond from the high bank of the River Slea.

Description and Management

Most of the reserve area is open water, probably created originally as a borrow-pit when the river bank was built. There is a reed fringe, with purple-loosestrife, greater spearwort, meadow-rue and both species of reedmace. Breeding birds include mute swan, Canada goose, mallard, moorhen, little grebe, tufted duck and reed and sedge warblers.

Little management is required other than the maintenance of a sluice to control the water-level, and the clearance of any dumped refuse.

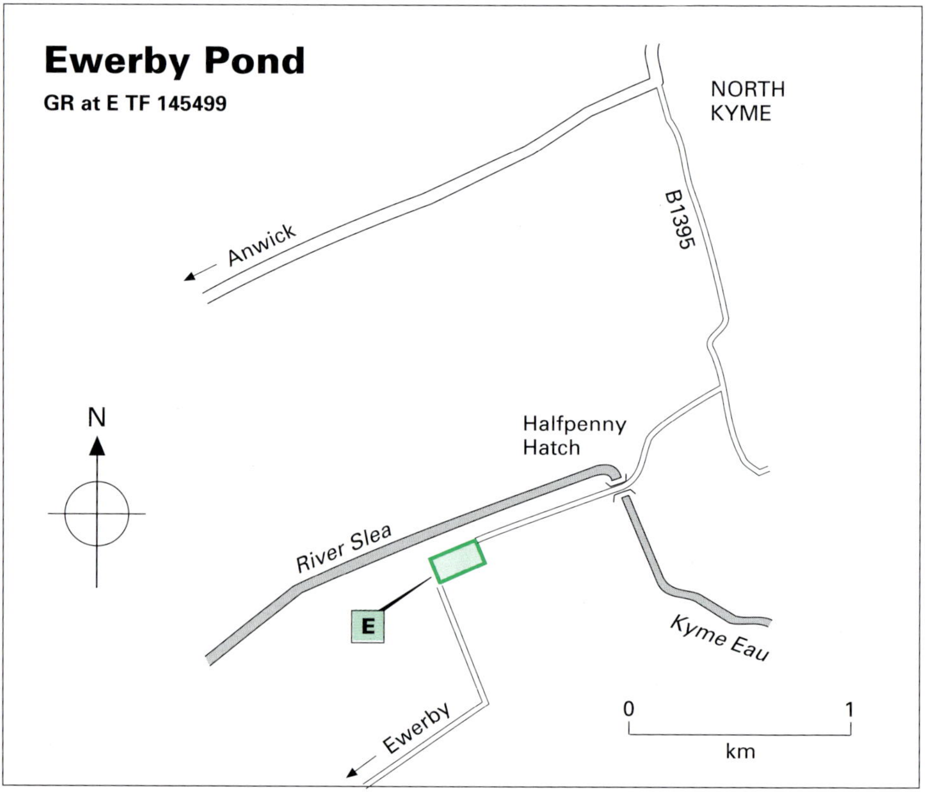

Fairfield Pit

Barrow-upon-Humber
OS: 112 GR: TA 080243 9.4 hectares (23.3 acres)
Part freehold, part management agreement 1987
Map ref: 85

Location and Access

The pit is situated close to the Humber Bank at New Holland. Pass over the level crossing at New Holland, then turn left along the track parallel to the railway line and follow it to the parking space on the Humber Bank.

Description and Management

Fairfield Pit is one of the Humber Bank clay pits, but its water conditions are more saline than the others. The pit is attractive to wading birds and surface-feeding wildfowl. During the autumn and winter many different bird species may be seen from the sea bank. Breeding birds include coot, moorhen, mallard, and reed and sedge warblers. Water-levels in the reserve are controlled to increase its attractiveness to wildfowl. In 1992 the Trust purchased the foreshore in front of the pit. The foreshore is a good place to observe both passage and wintering waders.

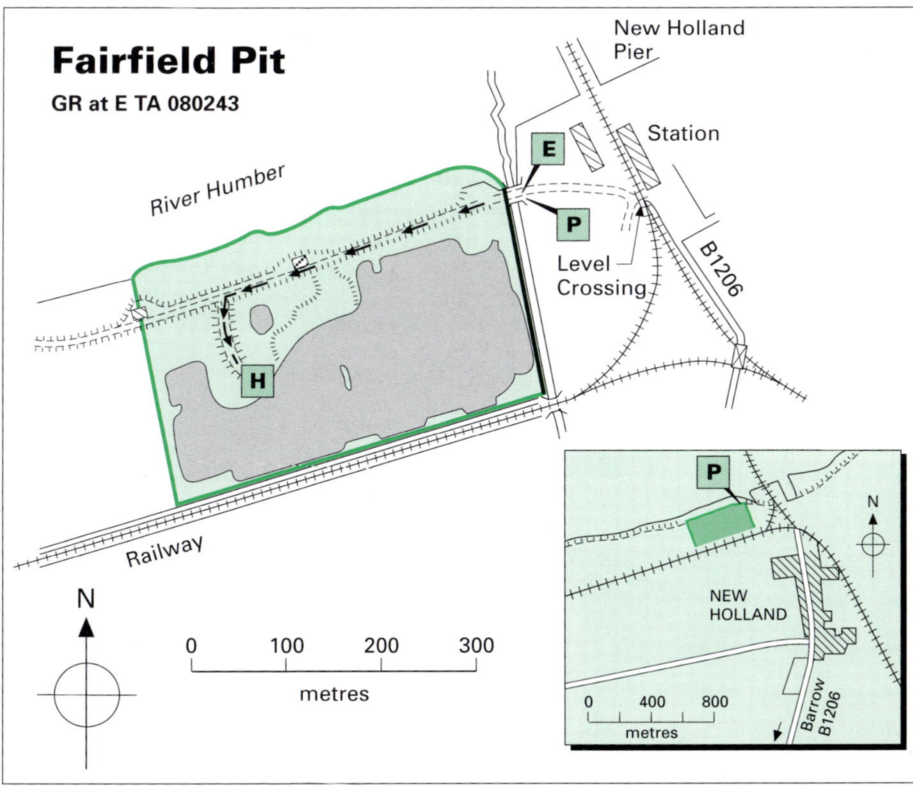

Far Ings (incorporating Barton Reedbed)

Barton-upon-Humber
OS: 112 GR: TA 011229 and TA 023230 58.6 hectares (144.9 acres)
The Sam van den Bos Memorial Reserve. Part leased, part freehold 1973
Map ref: 67 and 4

WC SSSI

Location and Access

The reserves adjoin Far Ings Lane and are a short distance to the west of the Humber Bridge. A well maintained visitors' route around the reserve can be accessed at the Visitor Centre in Far Ings Lane. The path is suitable for wheelchair access over most of its length (except when sheep are present when the gates have to be kept closed) and there is access for disabled persons at two bird hides. There are public toilets, including facilities for disabled persons, at the Visitor Centre. There are six hides on this route overlooking lakes, reedbeds and scrapes. The Barton Reedbed area of the reserve may be reached along the path on the river bank or along Far Ings Lane (see map). The two Westfield Lakes, to the east of the Visitor Centre, were purchased by the Trust in 1996 with the aid of a grant from the Heritage Lottery Fund. To avoid disturbance to birds and to sensitive areas visitors are asked to keep to waymarked or mown paths and to use the hides.

Description and Management

The chain of flooded clay pits and extensive reedbeds along the Humber Bank from west of Barton-upon-Humber to New Holland is a legacy of the brick and tile industry, which flourished there from the Middle Ages to the present century, and, more recently, of cement making.

The Far Ings and Barton Reedbed reserves comprise the open water of Ness Lake and Westfield Lakes, large areas of reedbed, recently made 'scrapes' with shallow water, grassland, hedgerow and scrub, and estuarine saltmarsh, which links the two parts of the reserve. The varied habitats in the reserve maintain a wide range of plants: reedbed and swamp species, including both species of reedmace and bur-reed; calcicoles, such as kidney vetch, wild carrot and mignonette on the river bank

The Humber Bridge from Ness Pit, Far Ings

Far Ings & Westfield Lakes

GR at E TA 011229

Barton Reedbed

GR at E TA 023230

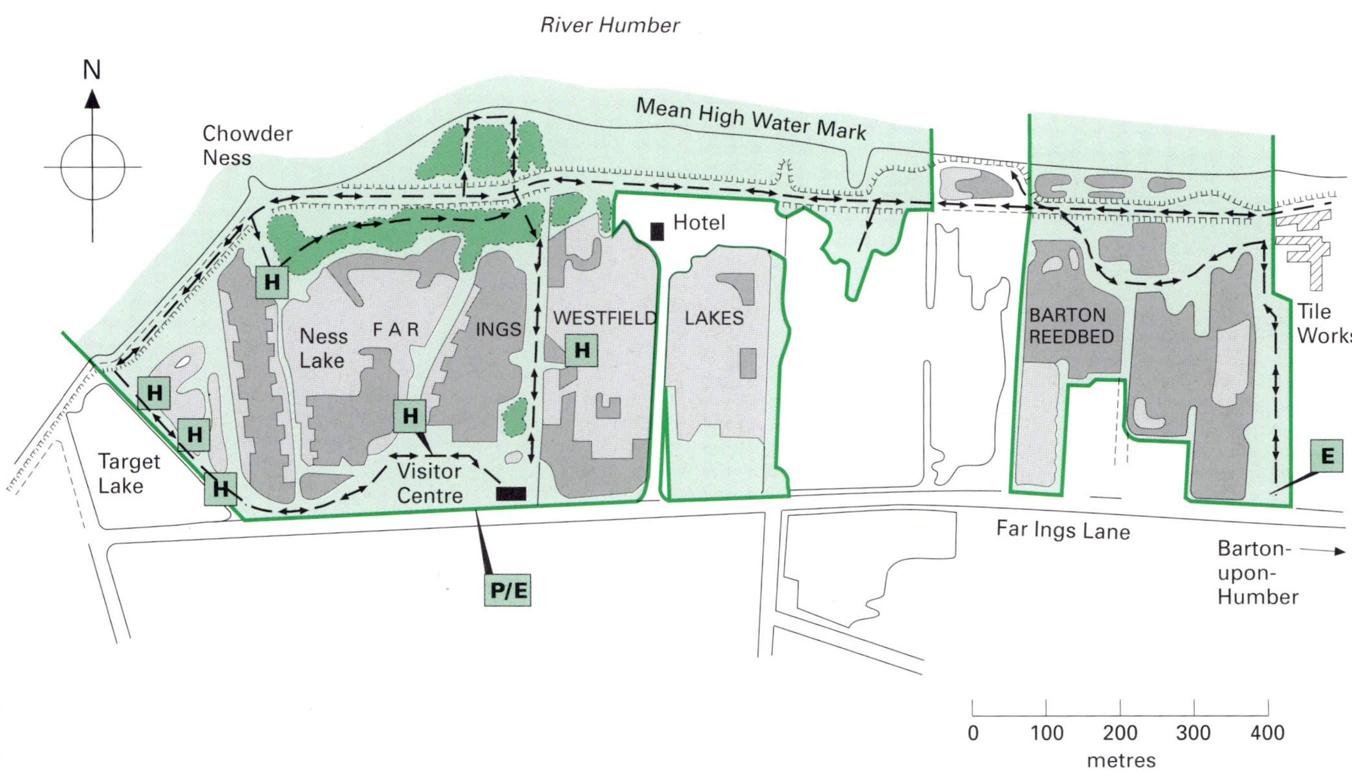

in the vicinity of the old cement works; and saltmarsh plants, such as sea aster and sea-milkwort. Butterflies include common blue, holly blue, gatekeeper and small copper, and more than 200 species of moths have been recorded, including a number of reedbed specialities. Birds are the outstanding attraction of the reserve. The Barton–Barrow clay pits are among the few breeding sites in Lincolnshire for the bearded tit, which often form conspicuous flocks in late summer; other notable nesting species include great crested grebe, water rail, shoveler, pochard and ruddy duck. Whitethroat, lesser whitethroat, blackcap, and garden, reed, sedge, grasshopper and willow warblers, and, occasionally, chiffchaff also breed on the reserve, the numbers of reed warblers being the most significant. The total number of warblers nesting in the reserve make Far Ings one of the most important British reserves for these summer residents. Scrapes, created in 1986, attract waders on passage, and common terns breed annually on rafts provided for them. In winter the open water and reedbeds are important for wildfowl. Numbers of duck can be large and include mallard, teal, gadwall, pochard, tufted and ruddy ducks, and shoveler, and occasionally rarities such as smew and red-crested pochard may be seen. Bitterns are sometimes seen in winter, and it is hoped that the extensive reedbed management currently being carried out will encourage these birds to return to breed as they did until 1980.

Management of the reserves consists of maintaining a wide variety of habitat for birds and insects, with reedbed management having a high priority. Reedbed management consists of harvesting the reed for thatching, burning areas to encourage the growth of better quality reed, and creating channels and pools to increase feeding margins for the reedbed inhabitants. The results of these measures are being monitored by annual censuses of key species.

Bittern

Fir Hill Quarry

Little Cawthorpe
OS: 122 GR: TF 362828 0.4 hectares (1.1 acres)
Freehold 1982
Map ref: 68

Location and Access

Fir Hill Quarry is a small chalk pit situated on the west side of the road between Little Cawthorpe and Muckton, some 4 km (2.5 miles) south of Louth. There is limited space for parking at the roadside. Please do not climb the scree slopes of loose chalk.

Description and Management

The reserve is on the line of the old marine cliff that formed the eastern edge of the chalk before the Ice Ages. There is a fine view over the Marsh to the sea from the road. The road is an extension of Barton Street, a prehistoric trackway that follows the edge of the chalk Wolds from the Humber to the Wash. The chalk pit has been quarried at two levels. The upper level is wooded with ash and wych elm, although most of the latter has succumbed to Dutch elm disease. The lower level has grassy banks, with salad burnet, marjoram, wild basil, pyramidal orchid, common spotted-orchid and other chalk plants. Like most flowery chalk pits it is a haven for most of the common butterflies.

Management consists largely of scrub control and mowing.

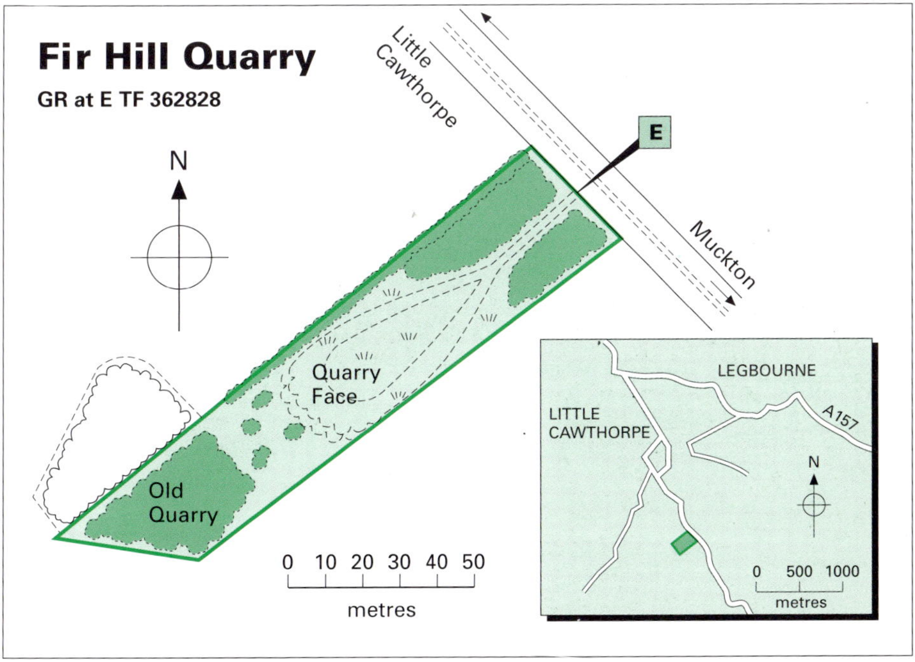

Frampton Marsh

Fishtoft and Wyberton
OS: 131 GR: TF 365393 (entrance TF 365393) 171.8 hectares (424.7 acres)
Freehold: 139.5 hectares (344.7 acres) 1976
Leasehold: 32.4 hectares (80 acres)
Map ref: 19

SSSI

Location and Access

Frampton Marsh lies between the outfalls of the River Witham (The Haven) and the River Welland. The major part of the reserve lies in the area known as The Scalp. It is reached via the marsh roads from either Frampton or Wyberton. Cars may be parked at the roadside at about grid reference TF 353396, from where the reserve is reached on foot by way of a track known as Commissioners' Drove. Alternatively, cars may be parked at about grid reference TF 363386; from this point proceed on foot to the sea bank and turn left. Care should be taken to check tides before venturing out onto the marshes as the creeks fill very quickly. Good views of the marsh and the river may be had by walking out along the Haven Bank.

Description and Management

The reserve is part of the most mature saltmarsh in the Wash and is exceptionally rich in plants, birds and invertebrates. The upper levels, which have extensive zones of sea-lavender, sea aster and sea-purslane, are intersected by large creeks, one of which was the old course of the River Witham before the new cut was made in 1880. The area supports a black-headed gull colony, while common tern, redshank, oystercatcher, reed bunting, meadow pipit and skylark are all regular breeders. In winter the saltings attract wigeon, mallard, shelduck, teal and brent geese, with large flocks of finches and buntings, notably linnet and twite. The tidal mudflats form part of the wader feeding grounds, which give the Wash its international status. Large flocks of dunlin occur, as well as considerable numbers of grey plover, whimbrel, curlew, bar-tailed godwit and greenshank.

Wildfowling rights are let to the South Lincolnshire Wildfowlers' Club who participate in the care and wardening of the reserve. Grazing by cattle is an important part of management. The Trust's reserve is grazed together with the RSPB's reserve to the south. Parts of the Kirton Marsh to the south are also subject to agreements between landowners and English Nature, so that all the saltmarshes between the Witham and Welland outfalls are under protective management. The whole of the Wash has been designated by the Government as a Ramsar Site and a Special Protection Area.

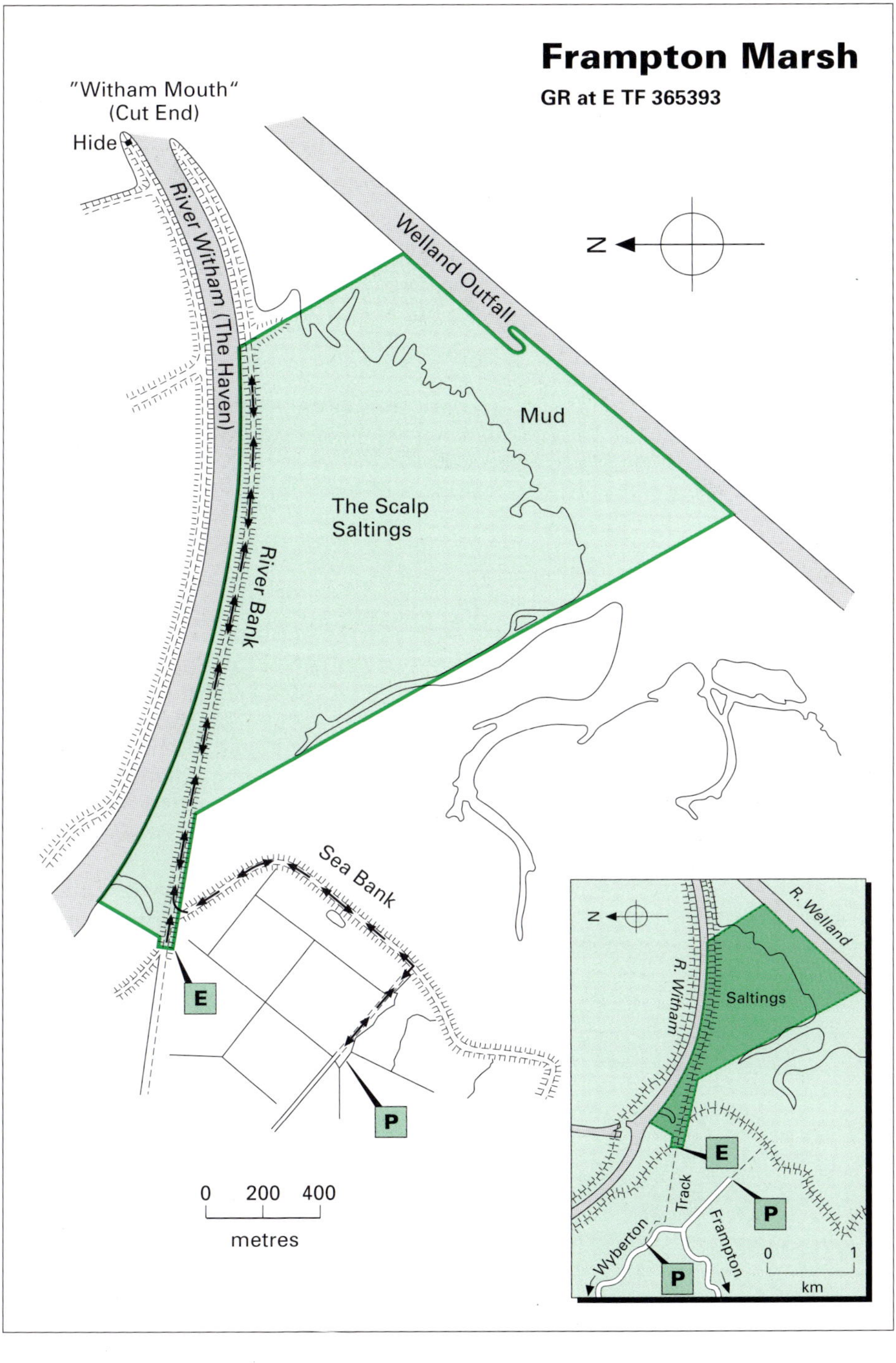

Frampton Marsh
GR at E TF 365393
"Witham Mouth"
(Cut End)
Hide
River Witham (The Haven)
Welland Outfall
N
Mud
The Scalp
Saltings
River Bank
Sea Bank
E
P
0 200 400
metres
R. Welland
R. Witham
Saltings
Track
Wyberton
Frampton
0 1
km

Freshney Bog

Grimsby
OS: 113   GR: TA 240092   14.2 hectares (35 acres)
Management agreement 1983
Map ref: 73

Location and Access

The reserve is situated on the outskirts of Grimsby on the A1136 road to Immingham. There is convenient parking off the A1136 next to St Michael's Church (TA 242090). Access to the site is immediately on the right after crossing the bridge over the River Freshney.

Description and Management

Freshney Bog was once famous as a rich botanical site, with plants such as grass of Parnassus and large bitter-cress, but canalisation of the River Freshney, drainage and refuse tipping schemes have destroyed much of the site and its wildlife interest. What remains is now part of North-East Lincolnshire Council's scheme, called the Freshney Parkway: an open space with walkways along the lower reaches of the Freshney eventually reaching the town centre and entering Grimsby docks. The Trust manages the Freshney Bog area as an urban conservation project involving the local community and schools.

The site is bounded to the south by the River Freshney and to the north by a drainage ditch, which contains much reed and water forget-me-not. There are sizeable areas of reed, and the rough grassland has a good display of cuckooflower in the spring and ragged-robin in the wetter flushes. Kestrels hunt over the grassland and skylarks nest. Several small ponds have been constructed where frogs and common newts abound. Over one thousand native trees have been planted, including ash, oak, silver birch and field maple. Areas have also been sown with seed from the Trust's Heath's Meadows, and plants, such as cowslip, that were rescued from development sites have been replanted.

Management involves the taking of a hay crop and maintaining paths. A mere has been constructed, and the hedge along the Great Coates Road has been laid.

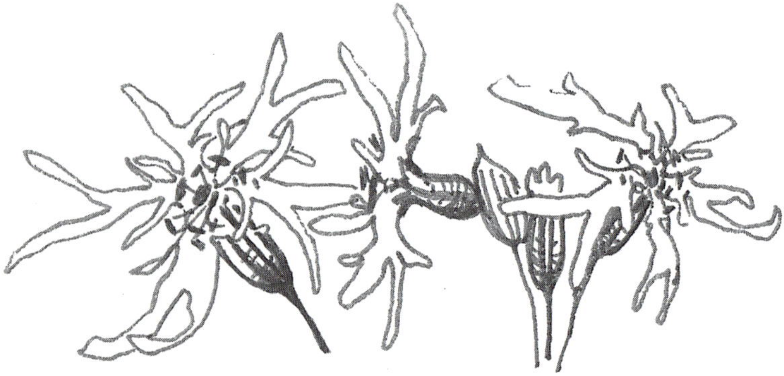

Ragged-robin

Freshney Bog

GR at E TA 240092

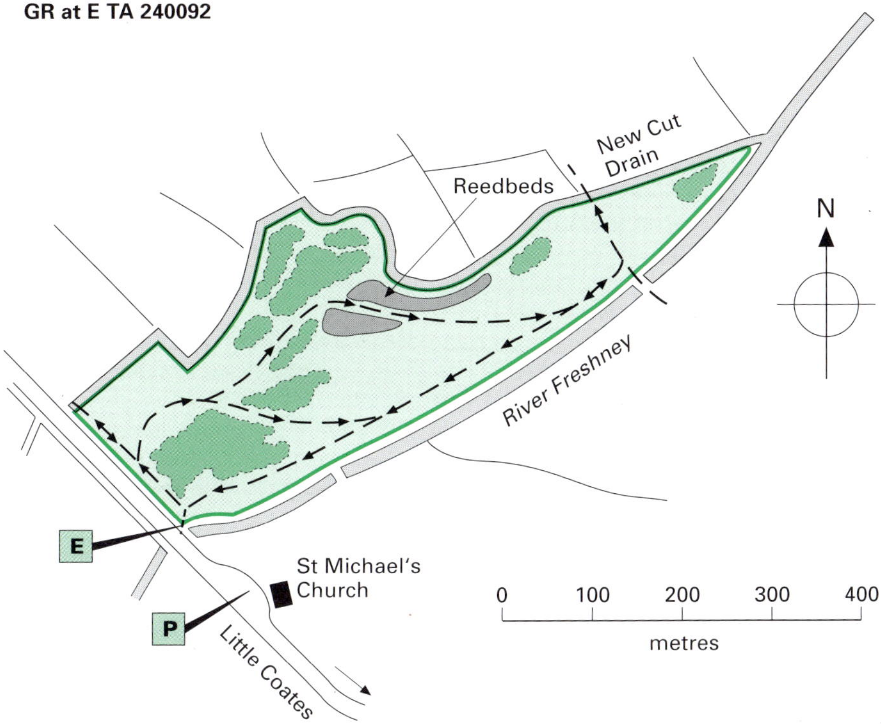

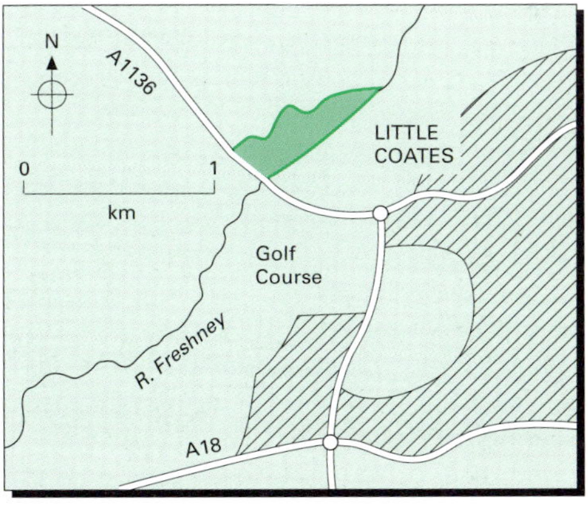

Friskney Decoy Wood

Friskney
OS: 122 GR: TF 474571 5.9 hectares (14.6 acres)
Freehold 1950. Gift of Mrs DK Booth
Map ref: 20

Location and Access

The Decoy Wood lies 1.5 km (0.9 miles) north of Friskney village on the north side of the Fodder Dyke Bank road from Eastville to Friskney. Access is via a footbridge some 500 m west of Marfleet's Bridge. Cars may be parked at the roadside. Cross the footbridge and follow the field edge path to the reserve entrance. The waymarked route is uneven and care should be taken to avoid exposed tree roots, rabbit holes, tree stumps and fallen branches.

Description and Management

The decoy, the most famous in the East Fen (the last great fen to be drained), survived with diminishing duck catches until 1878 when it finally went out of use. The fen had been a great reservoir for wildfowl but with the drainage begun in 1801 the decoy started to run down. Yet even in 1866 over one thousand ducks were taken, so we can only guess at the numbers caught in its heyday! The decoy was built by George Skelton (1760–1840) and was operated by his family until 1860.

Surrounding the old decoy pipes is woodland, which is on a thin peat soil, mostly consisting of birch, sycamore and Scots pine, with rowan, alder, willow and aspen. There are also larch, spruce and Corsican pine. Some woodland may have grown here before draining of the East Fen, but most of the trees have been planted.

The ground flora is dominated in parts by bracken and buckler-fern. The climbing corydalis, a localised plant in eastern England, occurs throughout the wood. Over 30 species of birds have bred, and large numbers use the wood in winter for feeding and roosting.

The main objectives of management are to maintain and enhance a varied woodland of mainly native trees and to retain the relict decoy pond and pipes as a wetland area.

Waymarked Route 2 km (1.2 miles)

Starting from the Fodder Dyke follow the path across the fields to the reserve entrance. The access path was made wide enough for a horse and cart to carry grain to 'feed-up' the decoy.

Just inside the reserve entrance is a small open area with a hut. Buddleia bushes grow here and in late summer their flowers attract many butterflies, mostly small tortoiseshell, peacock and red admiral. Follow the waymarked path indicated. As you enter the wood the peaty soil is dominated by bracken, but the climbing corydalis grows beneath the fronds. Its intricate creamy flowers repay closer examination. On the right of the path there are Scots pine, which provide feeding ground for goldcrest and various tits; to the left are some sycamores of unusually great size.

A little further on and to the north you will see farmland that is at a lower level than the reserve due to shrinkage caused by drainage, oxidation of the peat and cultivation. The peat of the reserve has also shrunk somewhat, leaving the roots of many trees exposed and vulnerable to windblow. The path turns sharply south and, further along, common buckler-fern takes over from bracken as the dominant ground cover. The path weaves through a stand of Norway spruce, where tawny owls are often heard or seen.

In the south-east corner of the reserve small nursery plots will be found where young trees are grown for planting out on the reserve.

The route then passes through an area with ash, birch, larch and alder. Look for redpoll, treecreeper and spotted flycatcher. In winter siskins may also be seen. The path ends at the open area by the entrance.

In addition to the main waymarked route, visitors may like to explore the path in the old south-east decoy pipe (see map). Proceed quietly in order not to disturb any birds by the pond: you may see a fishing heron, a kingfisher, mallard or other duck, though this will depend on the current water-level. With more intensive drainage of the surrounding farmland and the low rainfall of recent years, the water-level in the old decoy pond has dropped drastically. However, in March 1993 the pond was partly redug and restored. The greater area of water has resulted in an increase in waterfowl and in the populations of dragonfly species. In August and September migrant birds use the area, while in winter, just before dusk, flocks of thrushes drop in to roost.

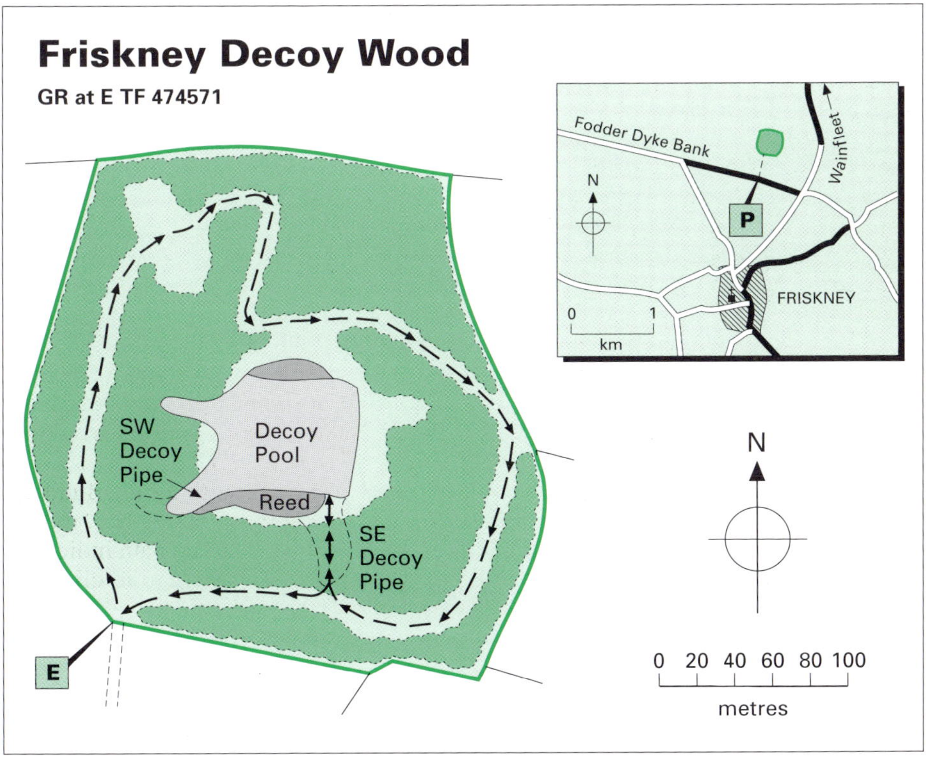

Furze Hill

Hagworthingham
OS: 122 GR: TF 346689 4.7 hectares (11.6 acres)
Freehold 1979
Map ref: 21

Location and Access

The reserve lies to the south of the village of Hagworthingham, about 6.4 km (4 miles) east of Horncastle. Turn off the A158 road and pass through the village centre, following Water Lane to a ford. The entrance to the reserve is on the left immediately after crossing the ford. There is limited parking space by the entrance. Furze Hill lies along the route that joins Snipe Dales to Upper Sow Dale. On the opposite side of Water Lane there is a footpath that leads to Snipe Dales. Sow Dale may be reached by following the path through Furze Hill (see page 177).

Description and Management

The reserve consists of hillside grassland and streamside habitats in the valley of a tributary of the River Lymn — the same brook that rises and flows through the Snipe Dales reserve. On entering, the first part of the reserve is a narrow meadow squeezed between the brook and an old hedgerow. The uneven surface of the ground reveals the former courses of the brook. Alongside the brook grow willows and alders where treecreepers and willow tits occur. This habitat is akin to that found in the alder carrs of the south-east Wolds, and such plants as moschatel and the opposite-leaved and rare alternate-leaved golden saxifrages are to be found. The main part of the reserve lies beyond this meadow on the opposite side of the stream over a wooden bridge (not the concrete one seen on the left soon after entering the reserve but about 100 m further on). There are attractive streamside habitats along the valley and hillside flushes, which are associated with the junction of the Spilsby Sandstone and the Kimmeridge Clay where many wet-loving plants thrive, including the uncommon blinks. The Kimmeridge Clay is exposed in places along the bank of the stream. Beside the stream is the site of an old water mill (marked by a nettle-covered mound). Nearby is a willow on which grow a remarkable variety of epiphytes, including ivy, mosses, lichens, hawthorn and gooseberry. On the north of Furze Hill is an old sand quarry with dense gorse cover, which provides good nesting sites for small birds, particularly linnets. This drier grassland supports heath bedstraw, dove's-foot crane's-bill, common stork's-bill, bird's-foot and parsley-piert.

Management is designed to maintain and improve floristic diversity by grazing at certain times of the year.

Furze Hill

GR at E TF 346689

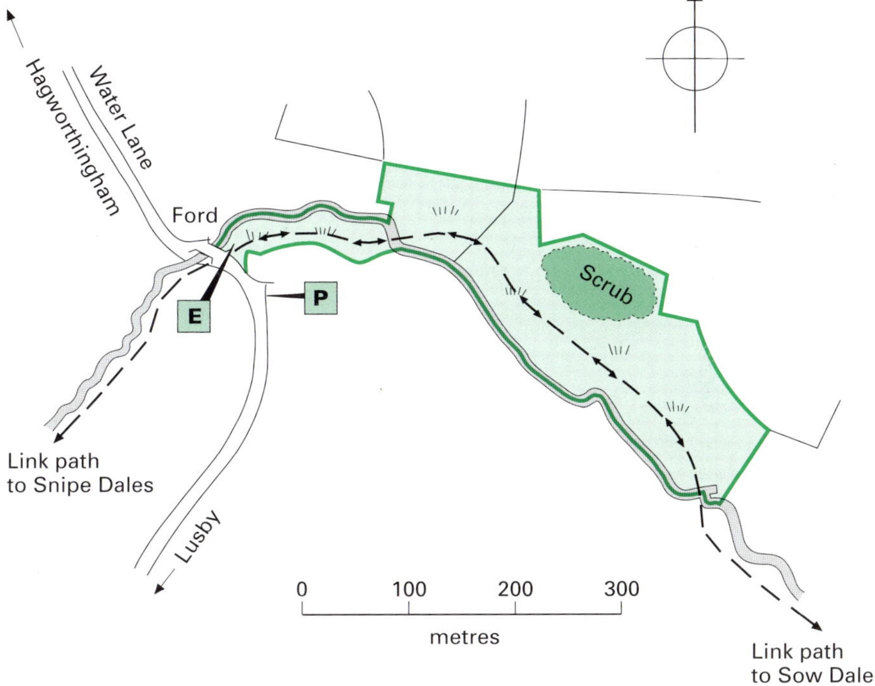

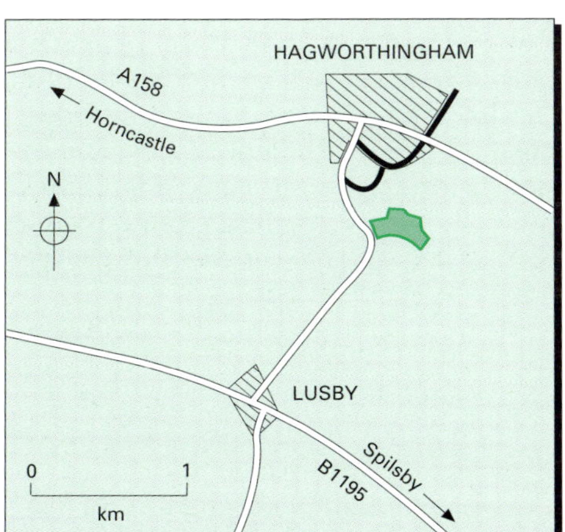

Gibraltar Point National Nature Reserve

Wainfleet St Mary and Croft
OS: 122 GR: TF 556581 (south car park) 437.4 hectares (1080.9 acres)
Part leasehold, part freehold 1949
Map ref: 22

The Gibraltar Point National Nature Reserve is managed by the Trust on long leases from Lincolnshire County Council (367 hectares: 907 acres) and from East Lindsey District Council (60.7 hectares: 150 acres). In addition, the Trust owns 9.7 hectares (23.9 acres) of the reserve known as Syke's Farm and Jackson's Marsh. The earliest declaration of part of the reserve was in 1949.

Location and Access

The reserve lies between Skegness and the estuary of the River Steeping, which rises in the Wolds as the River Lymn, often known as Wainfleet Haven at this point. The access points are at Seacroft Esplanade at the southern end of Skegness (TF 566610) and at two car parks on Gibraltar Road. The main car park is the largest and is close to the Visitor Centre and Field Station. A charge is made for the car parks all year round. Although this cannot be waived for members, it is usually possible to use the car park behind the Field Station without charge if a membership card is displayed in your windscreen. There is open public access to most of the reserve, but byelaws are in force. Certain sanctuary areas, which may vary from time to time, are designated, and all visitors are asked to respect these. The Visitor Centre is normally open every day of the week from May to October and at weekends during the winter. A *Walkabout Guide* is available from the Visitor Centre, where there are also extensive displays showing the history, habitats and wildlife of the reserve, and a wide range of gifts, books, stationery and educational items is on sale. The adjoining residential Field Station with its bird observatory is maintained by the Trust. A prospectus is obtainable on request from the Warden or the Trust's headquarters in Horncastle. The Wash View Point — a converted wartime gun emplacement — is situated 400 yards to the south of the Field Station and is open at all times. Members may also visit Syke's Farm, which is not open to the public. A small hide overlooks Lambert's Pond. From the main car park walk back along the road and, immediately after crossing the sea bank (The Hump), turn left through the gate. There is a waymarked route that returns to the road a short distance to the north. Many of the walkways are suitable for disabled persons, and both the Visitor Centre and the toilets have wheelchair access.

Description and Management

Within this reserve are the most extensive sand dunes on the Lincolnshire coast, and accretion of wave-built ridges roughly parallel to the coast is still proceeding rapidly. Between the ridges strips of saltmarsh are formed and are eventually reached only by the highest spring tides. From aerial photographs and maps this pattern of accretion can be traced back to the early part of the last century. The Western Dunes, alongside the public road, were in existence when the first 1" Ordnance Survey map was made in 1825, but they were then the frontal dunes. A

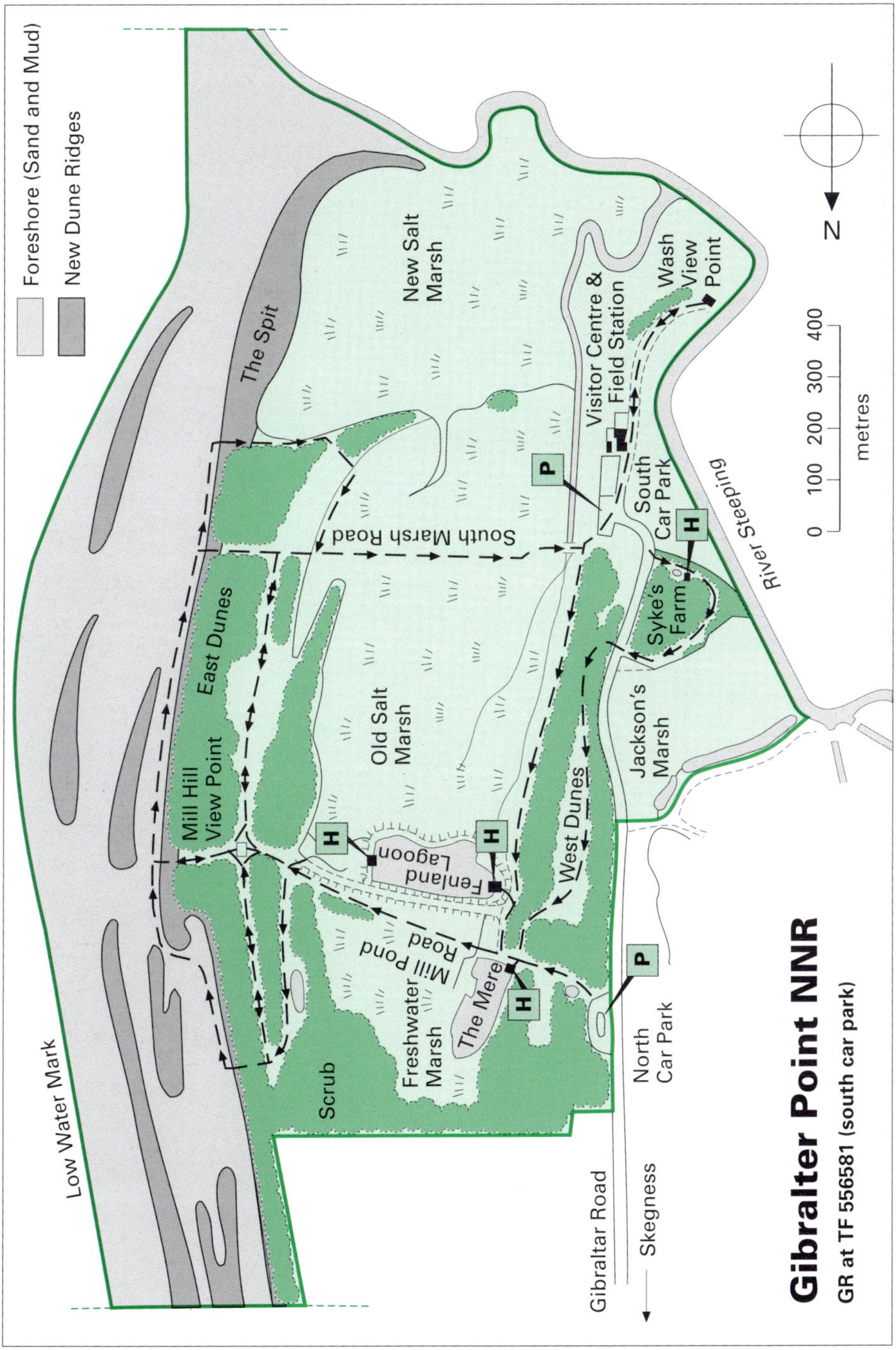

Foreshore (Sand and Mud)
New Dune Ridges
Low Water Mark
The Spit
New Salt Marsh
Visitor Centre & Field Station
Wash View Point
South Car Park
River Steeping
South Marsh Road
Syke's Farm
Jackson's Marsh
East Dunes
Old Salt Marsh
Mill Hill View Point
West Dunes
Fenland Lagoon
Mill Pond Road
The Mere
Freshwater Marsh
Scrub
North Car Park
Gibraltar Road
Skegness
P
H
N
0 100 200 300 400
metres
Gibralter Point NNR
GR at TF 556581 (south car park)

second ridge about 500 m seawards, known as the East Dunes, began to develop 100 years ago, and at the southern end a long spit extends towards the river outfall. In the shelter of the spit a new saltmarsh has developed during the last 30 years. This now extends as far as Wainfleet Haven. Beyond the East Dunes a further ridge is being built up causing more mud and saltmarsh to accrete. To the south, outside the reserve, some of the most extensive sand and mudflats and saltmarshes in Britain stretch southwards into the Wash.

All the stages in colonisation and stabilisation of sand and mud by plants can be seen in the reserve, and there are splendid examples of succession of vegetation. The mobile dunes support early colonisers, such as sea rocket, prickly saltwort, sand couch, marram and lyme grass. The older dunes are much more varied, with sand fescues, sand sedge and a variety of other plants, including bird's-foot trefoil, lady's bedstraw and dense scrub of sea-buckthorn, which is better represented here than anywhere else in Britain. On saltmarshes the first colonists of bare mud, such as glasswort (locally known as samphire) and sea-blite, give place to marshes dominated by sea meadow-grass, sea-purslane and sea-lavender. In areas less frequently reached by the tides there is sea wormwood, rock sea-lavender, sea-milkwort, sea-spurreys and other less common species. In the succession of both dune and saltmarsh rare plants occur. Among these on the newer dunes are sea-holly and sea bindweed, and at the edges of older dunes are slender centaury, knotted pearlwort and sea-heath. On the northern limit of the plant's range in Europe, a few specimens of shrubby sea-blite also occur.

North of the Mere is the Freshwater Marsh, which can be viewed from Mill Pond Road, where a wealth of wildflowers can be seen, including pyramidal orchids in profusion, and marsh-mallow.

The fauna of the reserve is exceedingly diverse. The open shore — an inhospitable looking habitat — has vast populations of invertebrate animals, such as lugworm, the peppery furrow-shell, the tiny spire shell *Hydrobia ulvae*, and the little shrimp-like *Corophium.* On the dunes are insects of many kinds. Resident butterflies include the uncommon green hairstreak, and there are many moths of seaside habitats, such as the coast and sand darts and the scarce pug. Hunting wasps, robber flies and many kinds of spiders add to the variety. Among larger animals, water shrew, harvest mouse and yellow-necked mouse have been recorded, and amphibians are present in the freshwater pools north of Mill Pond Road. The Natterjack toad has recently been successfully re-introduced after an absence of around 80 years.

It is the birds, however, that are the most conspicuous animals of the reserve. The sandy foreshore provides one of the few regular nesting sites on the Lincolnshire coast for the little tern. There are about 50 pairs of ringed plover, and smaller numbers of redshank and shelduck. The marshes and sea buckthorn scrub hold a high breeding population of small passerines: surveys have shown high densities of 35 species. Migration brings great numbers of passage migrants, including many rarities, while in winter brent goose, snow bunting, twite, fieldfare, redwing and other northern immigrants feed in the reserve. The sand spits and dune ridges provide a high tide roost for many thousands of waders that fly in from the Wash to await the ebb. The freshwater Mere and the 4-acre brackish scrape, called Fenland Lagoon, have been created to provide additional bird habitats. There is a viewing hide at the Mere and two hides at the Fenland Lagoon. These are most easily reached along the Mill Pond Road from the North Car Park.

The principal management problem of the reserve is to reconcile its intensive use for education and public enjoyment with the conservation of its wealth of plants and animals and natural features. Educational use is carefully planned, public access is regulated by bye-laws, car parks are provided adjoining the public road and tracks are well made and defined. There are full-time and voluntary wardens, and one of the main purposes of the interpretive programme is to enlist the interest and cooperation of visitors in conservation measures. The effects of human pressure on the flora and fauna of the reserve are constantly monitored. The freshwater marsh and parts of the dunes are grazed by cattle and sheep. Recent habitat improvements have included the dredging of ponds and borrow-pits and the clearing of buckthorn scrub in order to extend areas of dune grassland and dune slack. The 20-acre field adjoining Syke's Farm, known as Jackson's Marsh, has recently been purchased by the Trust. It is planned to create new wetland habitat here over the next few years.

Fieldfare, Gibraltar Point National Nature Reserve

Goslings Corner Wood

Langton-by-Wragby
OS: 121 GR: TF 144752 10.2 hectares (25.2 acre)
The George Leachman Memorial Reserve. Freehold 1987
Map ref: 79

Location and Access

The reserve is situated on the north side of the minor road known as Hoop Lane, which joins the B1202 Wragby–Bardney road with the A158 Lincoln–Skegness road. Access is across a grass field on the west side of Hoop Lane Farm. Cars may be parked on the verge of the narrow road, but care must be taken so as not to obstruct traffic. Inside the wood overgrown rides have been re-opened to provide a circular route. The route is unsuitable for wheelchairs.

Description and Management

Goslings Corner is one of the smaller Lincolnshire oak/lime woods, but it is an excellent representative of a distinctive type of these nationally important woodlands, and none has a richer flora. The ancient wood bank along the western and southern boundaries formed the edges of the former Langton Wood, which was largely destroyed between 1850 and 1870 and of which Goslings Corner is a remnant. The wood may also have been joined in relatively recent years to Great Scrubbs Wood, which lies, together with Little Scrubbs Meadow reserve, about half a mile to the south. The variety of soils in the reserve — ranging from mainly neutral to mainly acid — produces the variety of flora. The tree and shrub structure consists of scattered oak standards over coppice of small-leaved lime, oak, hazel, birch and field maple. Aspen, sallow, crab apple, dogwood, spindle and midland hawthorn also occur. The ground flora is impressively rich with abundant spring flowers, such as bluebell, wood anenome and woodruff, wood buttercup, yellow archangel, yellow pimpernel, ramsons and common valerian. On the more acid soils wood-sorrel and lily-of-the-valley occur, while rarer woodland plants include greater butterfly-orchid and bird's-nest orchid, and the attractive grasses wood melick and wood millet. Birds include great spotted woodpecker, nightingale, treecreeper, blackcap and other woodland warblers.

A series of coppice plots have been created off the main rides. The oldest plot at the southern end of the main ride was cut in the winter of 1989/90 and the last one in this present series is due to be completed near the turn of the century. The plots are covered in spring flowers for the first few years before the trees and shrubs re-grow and the leafy canopy closes in again. The plots also produce a varied age structure of trees and shrubs along the edge of the rides.

Goslings Corner Wood

GR at E TF144752

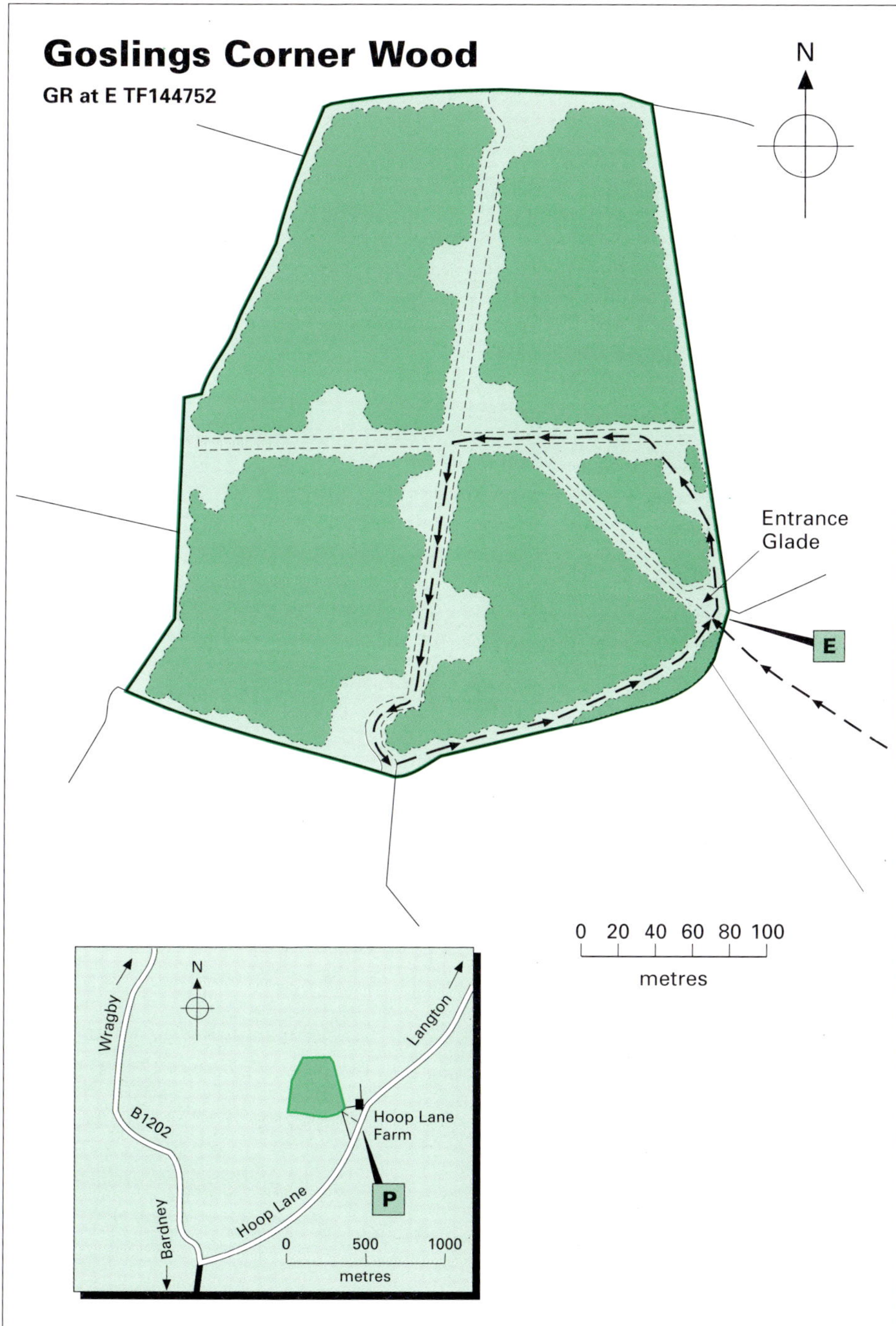

Great Casterton Road Banks

Stamford
OS: 130 GR: TF 006084 0.25 hectares (0.6 acres)
Freehold 1968
Map ref: 64

SSSI

Location and Access

The reserve is situated just west of Stamford on the south of the Old Great North Road and to the west of the junction with Sidney Farm Lane. Access to the site is difficult. It is best to park on Sidney Farm Lane well away from the junction and walk back to the Old North Road, and then west for some 5 m. The entrance is located to the right of the mature ash tree.

Description and Management

The area consists of a steep limestone bank rising from the main road, together with a buffer strip acquired in 1991. The top of the bank is fairly flat with a few hills and hollows. The site is renowned for its limestone flowers and abundant butterflies. The grassland contains typical limestone plants, such as wild thyme, common rock-rose, salad burnet, kidney vetch and quaking-grass. More unusual plants include sulphur clover, greater broomrape and perennial flax.

Ecological studies of insects associated with rock-rose have been carried out on the reserve. Despite encroachment of housing and improvements to the adjoining roads this small site retains much of its interest. Management includes the control of scrub and grassland mowing.

(Left to right) Rock-rose, kidney vetch, salad burnet, quaking-grass (above), wild thyme (below)

Great Casterton Road Banks

GR at E TF 006084

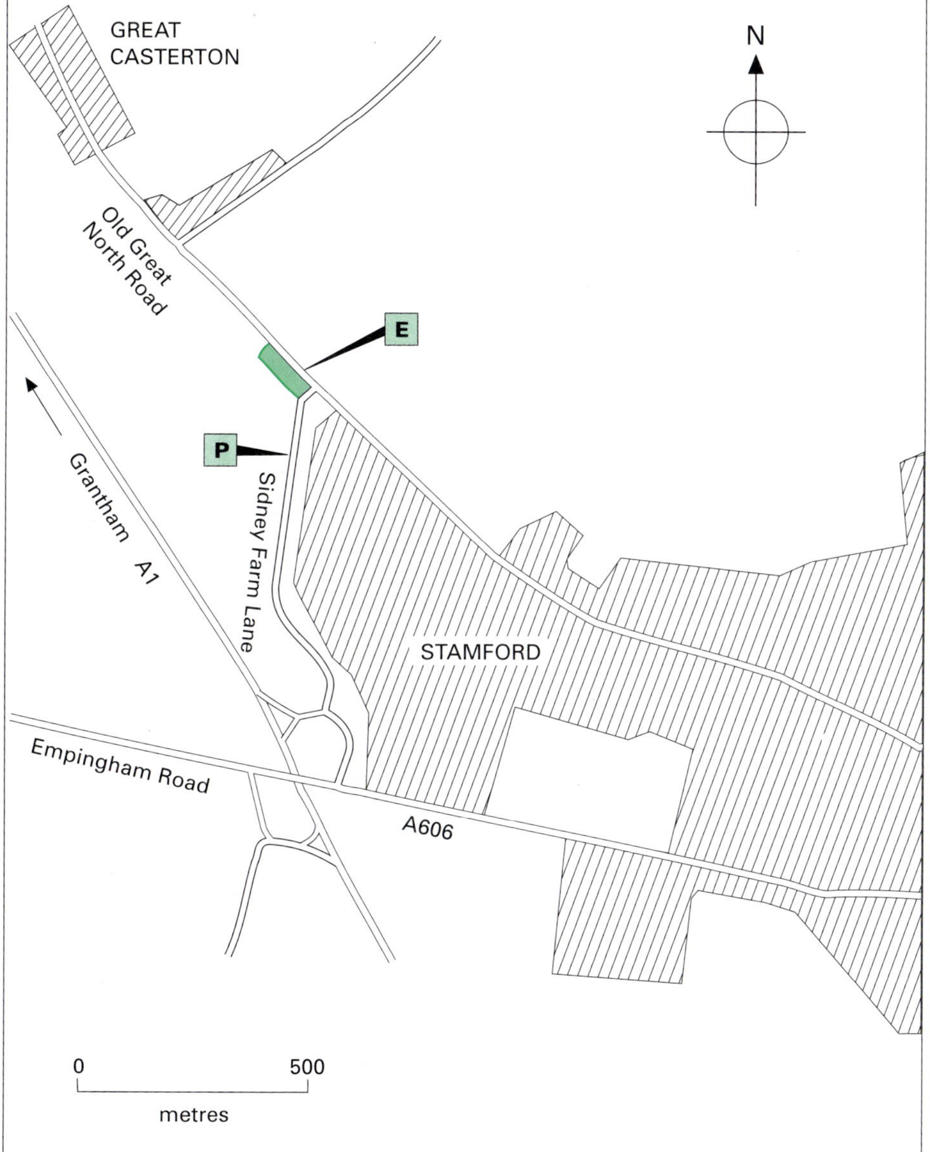

Hatton Meadows

Hatton
OS:121 GR: TF 167754 4 hectares (9.8 acres)
The Edward Mason Memorial Reserve. Freehold 1987
Map ref: 83

Location and Access

Situated on the north side of Hatton Wood, access is from the Wragby–Horncastle A158 road and via a hard track, which, if travelling from Horncastle, is on the left shortly after the Midge Inn and opposite the entrance to the British Gas site. Cars may be parked at the side of the track (see map). It is about a kilometre (0.6 miles) walk to the reserve. The hard track turns into a green lane shortly before the meadows. Please keep to the mown path before the hay is cropped. A circular walk can be made by using the two footbridges.

Description and Management

The reserve consists of two small damp meadows adjoining Hatton Wood (SSSI). Tall hedges surround most of the meadows and contain both species of hawthorn as well as crab apple, field maple, blackthorn, purging buckthorn and a few large oak trees. A small beck with a gravel bed makes a dog-leg course through the meadows, and there are two small ponds in the west field in which common water-plantain, yellow flag and water-crowfoot occur. Great burnet is common in the meadows and there is much meadowsweet in the wetter areas. Greater and lesser pond-sedge, lesser stitchwort, common spotted-orchid, greater burnet-saxifrage, sneezewort, cowslip, ragged-robin, yellow rattle, creeping-Jenny and lady's bedstraw are also present. Along the edge of Hatton Wood, woodland species, such as primrose, giant bellflower and wood avens, can be found, and angelica, valerian and water avens can be seen on the sides of the ditches. The tall hedges and proximity of Hatton Wood make the site attractive to birds with many breeding on the reserve, including garden and willow warblers, blackcap and nightingale. Many butterflies, including some scarce species, have been recorded, among which are purple and brown hairstreaks, white admiral, common blue and large skipper.

A return to the traditional management of hay-cutting and aftermath grazing with sheep since acquisition has allowed the sward to become more open, thus improving both the variety and quantity of herbs. Management work includes stock fencing, and hedge and ditch maintenance. The adjoining Hatton Wood is managed by the Forestry Commission and is open to the public. Visitors may enter the ride system from the reserve.

Hatton Meadows

GR at E TF 167754

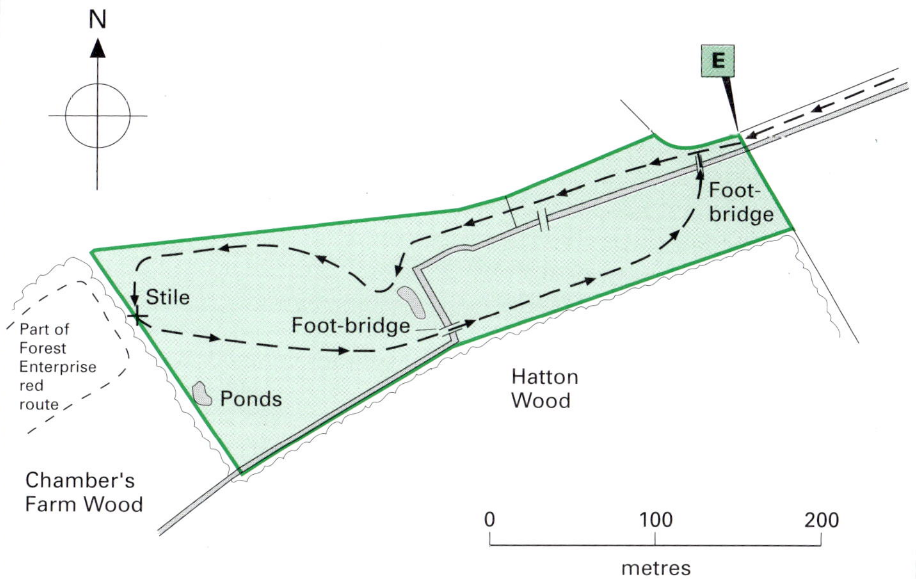

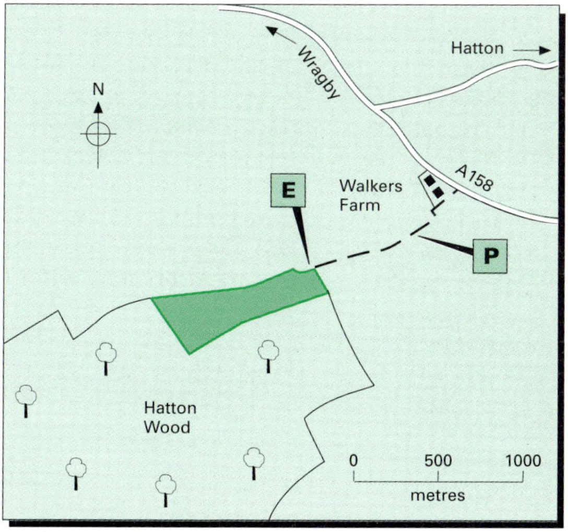

Haxey Turbary

Haxey
OS: 112 GR: SE 755017 14.7 hectares (36.3 acres)
Freehold 1958
Map ref: 23

SSSI

Location and Access

The reserve lies about 2 km (1.2 miles) to the north-west of Haxey village in the Isle of Axholme. At the western end of the village is a green triangle with an old stone cross ('Cross of Piety'). From here take the road running north signposted to Epworth. After 150 m this becomes Nooking Road; turn left into Haxey Carr Lane. Follow this past Haslam's Farm and Skyer's Farm. The road is rough and unmade. Ignore tracks to the left and right, and after about 2 km a reserve sign will be seen on the left. This area, known as Haxey Carr, is detached from the main part of the reserve. A small car park is reached further along the track. It is advisable to keep to the waymarked route. Care should be taken due to the uneven nature of the old peat workings. There is a public footpath on the main track through the reserve.

Description and Management

The reserve is a relict of bog that has been extensively dug for peat and subsequently colonised by birch. Interesting wet plant communities occur, and fen-sedge and the royal fern are still to be seen. Although the dense cover provides good habitat for some nesting birds, including grasshopper warbler and nightingale, there is a need to keep some open areas for other birds and to encourage a more diverse peatland flora. Birch thinning and localised clearance is carried out and a large glade will be found along the waymarked route. The glade contains an interesting mix of cross-leaved heath and fen-sedge. The deep holes are old peat cuttings.

Waymarked Route 1 km (0.6 miles)

From the car park follow the track in a westerly direction, which leads into the path that runs along the length of the reserve. Note the ridges and hollows that remain from peat excavations. Some of the hollows still retain bog plants, such as *Sphagnum* mosses and cross-leaved heath. Where birch has been cleared, good areas of heathland are to be found. Look out also for the royal fern. The twisting track eventually leads to a straight, raised trackway, which was probably constructed for peat extraction. Turn right and follow this back to the car park. In summer listen out for nightingales. Other birds that may be seen are woodcock, redpoll and willow tit.

Royal fern

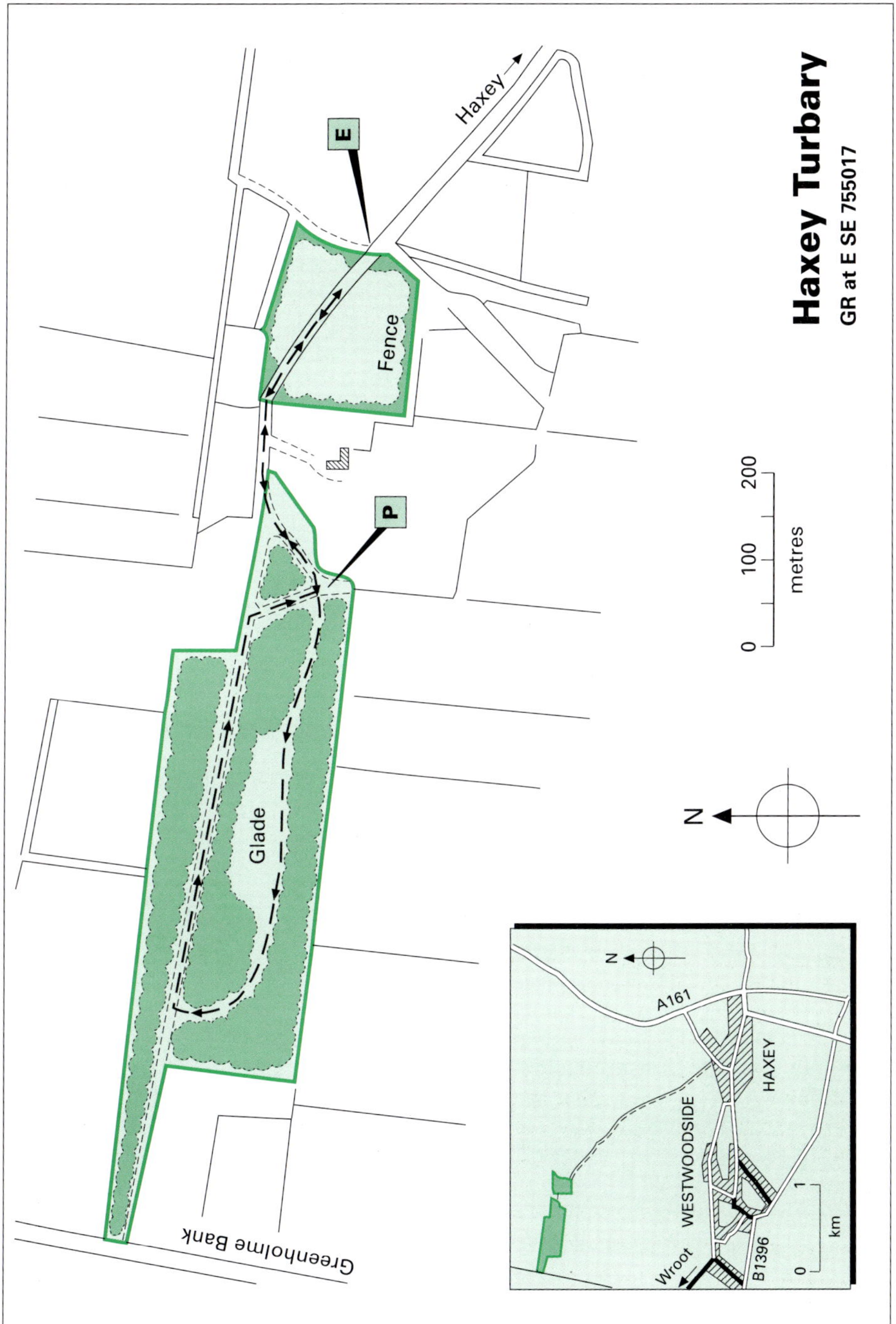

Haxey Turbary
GR at E SE 755017
E
P
Haxey
Fence
Glade
Greenholme Bank
0
100
200
metres
N
A161
HAXEY
WESTWOODSIDE
B1396
Wroot
0
1
km

Heath's Meadows

Bratoft
OS: 122 GR: TF 484640 (north entrance) 5.5 hectares (13.5 acres)
The Nurse Winifred Heath Memorial Reserve. Freehold 1969
Map ref: 24

Location and Access

The meadows are situated some 2 km (1.25 miles) south-west of Burgh-le-Marsh at Bratoft Ings. Access is from Ings Lane via a gate that leads into the first small corner meadow. There is room at the roadside for parking one vehicle. An alternative access (suitable for wheelchairs) is available by way of a footpath leading off Ings Lane on the north side of a small bungalow. Please keep close to the hedgerow through the access field, which does not belong to the Trust.

Research into the ecology of ancient meadows is carried out at the reserve by English Nature in cooperation with the Trust. Special experimental areas are pegged out during part of the year and these should not be entered.

The reserve is named after Nurse Heath of Nettleham in recognition of her services to the Trust. A memorial stone by the sculptor Richard Devereux was placed on the reserve in 1991.

Visitors are requested to keep to the edges of the fields and avoid trampling as much as possible, especially before hay time.

Description and Management

Bratoft Ings (an old Scandinavian word meaning meadow or pasture) are still divided into small fields, mainly meadow or pasture, with big hedges containing oak and willow trees. Each of the Trust's meadows has somewhat differing plant communities reflecting past differences of grazing, mowing and manuring treatment. They contain a wealth of grassland flowers: cowslip, the now rare green-winged orchid, spotted-orchid, adder's-tongue, great burnet, saw-wort, dyer's greenweed and devil's-bit scabious. This rich flora is maintained by continuing the traditional management of hay cropping and grazing. The hay is sold to a local farmer and the meadows are let for grazing from August to November. The Trust also owns some adjoining pasture, which is grazed by cattle during the summer months.

Green-winged orchid

Heath's Meadows

GR at E TF 484640 (north entrance)

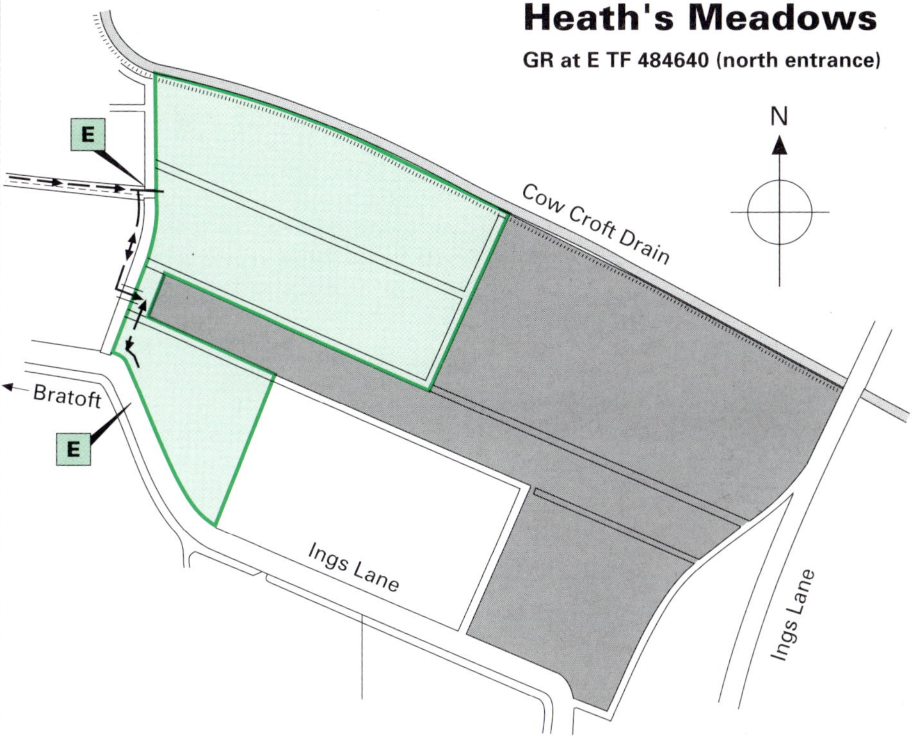

Other land owned by LTNC
but no access at present for members

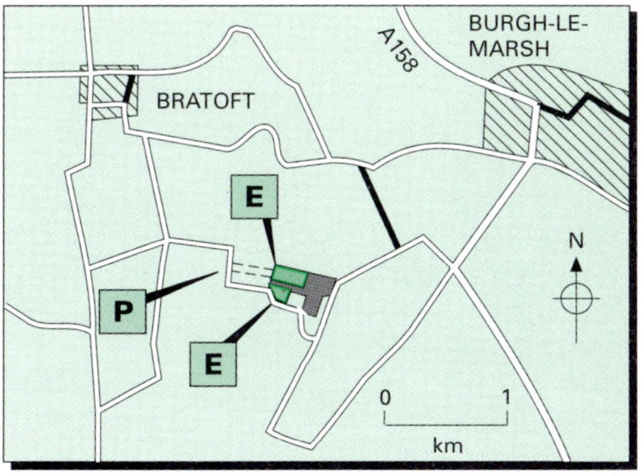

Hobhole Bank

Fishtoft
OS: 131 GR: TF 368416 5 hectares (12.4 acres)
Management agreement 1988
Map ref: 89

Location and Access

The reserve lies on the western bank of the Hobhole Drain close to its junction with the Boston Haven south of Fishtoft. The entrance is from Nunn's Bridge, from where public footpaths along both banks of the Hobhole Bank can be accessed. Reserve access is via stiles at either end and therefore there is no access for wheelchairs. The reserve is about 1 km (0.6 miles) in length.

Description and Management

The western bank has tall scattered scrub mainly of hawthorn but also some elder, willow and rose with a few sycamore. This scrub is attractive to a variety of birds and many breed here, including sedge warbler, common and lesser whitethroats, turtle dove, chaffinch, yellowhammer, linnet and greenfinch. In winter large numbers of finches and thrushes roost in the thickets as do long-eared owls. Birds to be seen on the drain include little grebe, heron, mallard, teal, tufted duck, pochard, wigeon, shoveler and goldeneye. Among the mainly coarse grasses, teasel, common knapweed, lady's bedstraw, yarrow, spiny restharrow and agrimony may be seen. In spring the abundant cowslips make an attractive display. On a calm summer's day many of our commoner butterflies may be seen.

The area is lightly grazed and rotational cutting of small areas of scrub will maintain the grassland and scrub mosaic.

Long-eared owls

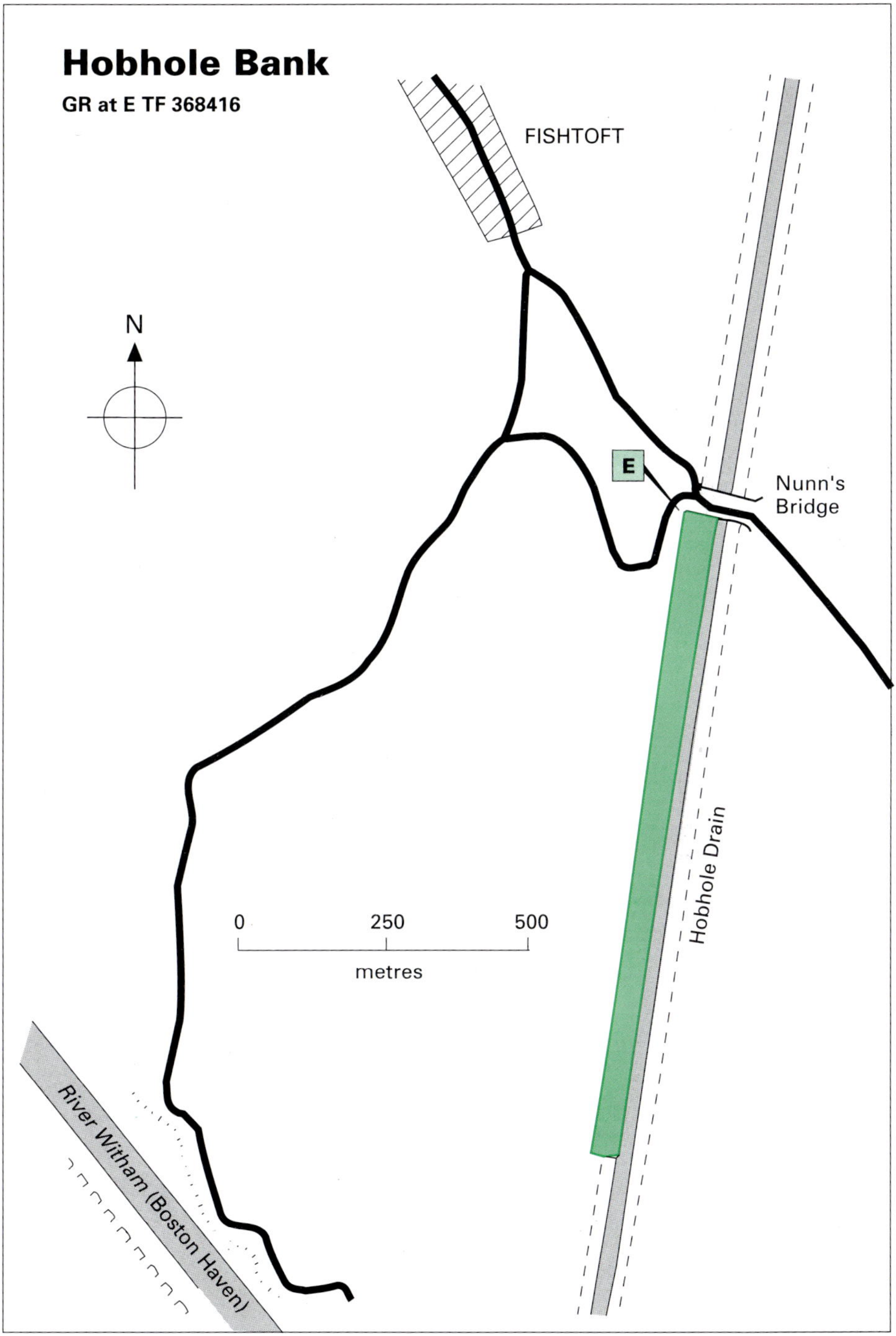
Hobhole Bank
GR at E TF 368416
FISHTOFT
N
E
Nunn's Bridge
Hobhole Drain
0
250
500
metres
River Witham (Boston Haven)

Hoplands Wood

Claxby-by-Alford
OS: 122 GR: TF 459718 14.2 hectares (35 acres)
Freehold 1964
Map ref: 25

Location and Access

Hoplands Wood lies on the south side of the Willoughby–Claxby road, about 4 km (2.5 miles) south of Alford and 1 km (0.6 miles) west of the B1196 road. The entrance gate is beside the road, where there is a space for cars on the verge. Visitors are requested to keep to the rides. A public footpath crosses the reserve on the main east-west ride. For a location plan with adjoining reserves see page 51.

Description and Management

An example of the once extensive oak/ash woods of the calcareous Boulder Clay soils of the Middle Marsh of east Lincolnshire, it lies just below the eastern edge of the Wolds and partly in the valley of a little chalk stream, Burlands Beck, which forms the southern boundary of the reserve and rises in the Claxby Spring reserve.

Oak and ash are the dominant trees, and there are fine standards of both species. Downy birch and alder occur throughout. Wych elm and English elm are also present but both have suffered from Dutch elm disease. The shrub layer includes hazel, field maple, sallow, field rose, dogwood, guelder-rose, hawthorn, blackthorn, honeysuckle and occasional spindle. High forest conditions pre-

Hoplands Wood

Hoplands Wood
GR at E TF 459718
N
P
E
Willoughby
Burlands Beck
0
50
100
metres

dominate in the eastern half of the wood, while elsewhere there are extensive areas of hazel coppice.

The moisture retentive clays and the closeness of the canopy make Hoplands a damp wood in which ferns and mosses abound. Primrose, sweet woodruff, moschatel and early-purple orchid are common, with bluebells in drier areas, and there are patches of herb-Paris. The rides have bugle, creeping-Jenny, wood and water avens, marsh thistle, lady's mantle, meadowsweet, spotted-orchid and twayblade.

There is a varied bird population at all seasons. Breeding species include woodcock, tawny owl, treecreeper, great and, occasionally, lesser spotted woodpecker and nuthatch, five species of tit and six species of summer warbler. Barn owls occasionally hunt in the rides, and redpolls and siskins visit birches and alders for seed in winter. Badgers frequent the wood from a nearby sett. A pond in the glade attracts spawning frogs, which abound in the wood.

In order to encourage a variety of flora and fauna, the main aims of management are: to perpetuate the high forest areas by thinning and by planting native tree species, notably oak; to restore the traditional coppice-with-standards system in appropriate areas; and to manage existing rides and glades, open up old ones and maintain the stream as an attractive feature. Considerable progress has been made in the achievement of these aims.

Early-purple orchids

Horbling Line

Horbling
OS: 130 GR: TF 115350 1.2 hectares (3 acres)
Management agreement 1985
Map ref: 72

Location and Access

Access is from the old railway bridge on the lane running westwards from the village of Horbling. There is a path through the northern section but access is not possible to the south.

Description and Management

The reserve section of this old railway track consists of a cutting and an embankment. Blackthorn and hawthorn scrub provide nest sites for whitethroat and lesser whitethroat, spotted flycatcher, nightingale and tree sparrow. In winter seed supplies attract flocks of finches, and there is a large thrush roost. The commoner butterflies are also abundant. Control of scrub and maintenance of open grassy areas are the main management aims.

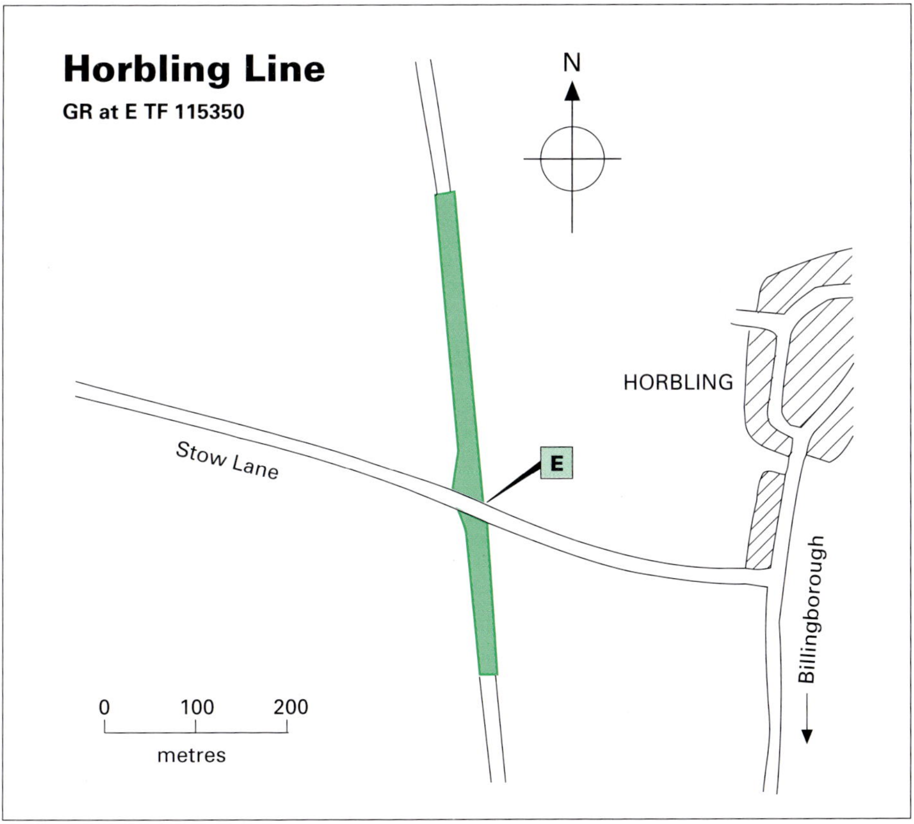

Huttoft Bank Pit

Huttoft
OS: 122 GR: TF 536793 4 hectares (10.1 acres)
Freehold 1959
Map ref: 44

Location and Access

The pit is situated on the west side of the sea bank road, about 1.6 km (1 mile) south of Sandilands. There is parking further south at the Huttoft Car Terrace, from where it is about a 0.8-km (0.5-mile) walk to the reserve. There is also room for two cars on the verge near the reserve entrance (opposite the public footpath sign on the sea bank). Cross the footbridge over the dyke and follow the fieldside track to the a hide situated on the bank. Visitors are advised that they should not walk round the pit. The track is unsuitable for people with walking disabilities.

Description and Management

This is the largest of the Sea Bank Clay Pits, which provided clay for sea bank repairs after the 1953 floods. There is a large open water area as well as extensive reedbeds. Emergent and aquatic vegetation includes species such as fennel pondweed, sea club-rush, great reedmace, brackish water-crowfoot, soft hornwort and yellow flag.

Moorhen, water rail, reed bunting, reed and sedge warblers, and bearded tit all nest on the reserve. Fifteen species of duck have been recorded, mainly during the winter, and Bewick's swan, slavonian, great crested, little and red-necked grebes, and red-throated diver have all been recorded. Rare migrants have included spoonbill, great reed warbler, little bittern and many others. Wading birds occur, especially in autumn, including common, green and curlew sandpipers, redshank and greenshank, and black-tailed godwit. Scarce visitors have included lesser yellowlegs and pectoral sandpiper. Herons are occasional visitors, and flocks of gulls also occur occasionally, including the Mediterranean gull. Various terns occur on passage. Short-eared owls occur in winter and marsh harriers in spring.

Management includes planting willow screens, cutting selected areas of reed and the creation of a new hedge alongside the access track.

Green sandpiper

Huttoft Bank Pit

GR at E TF 536793

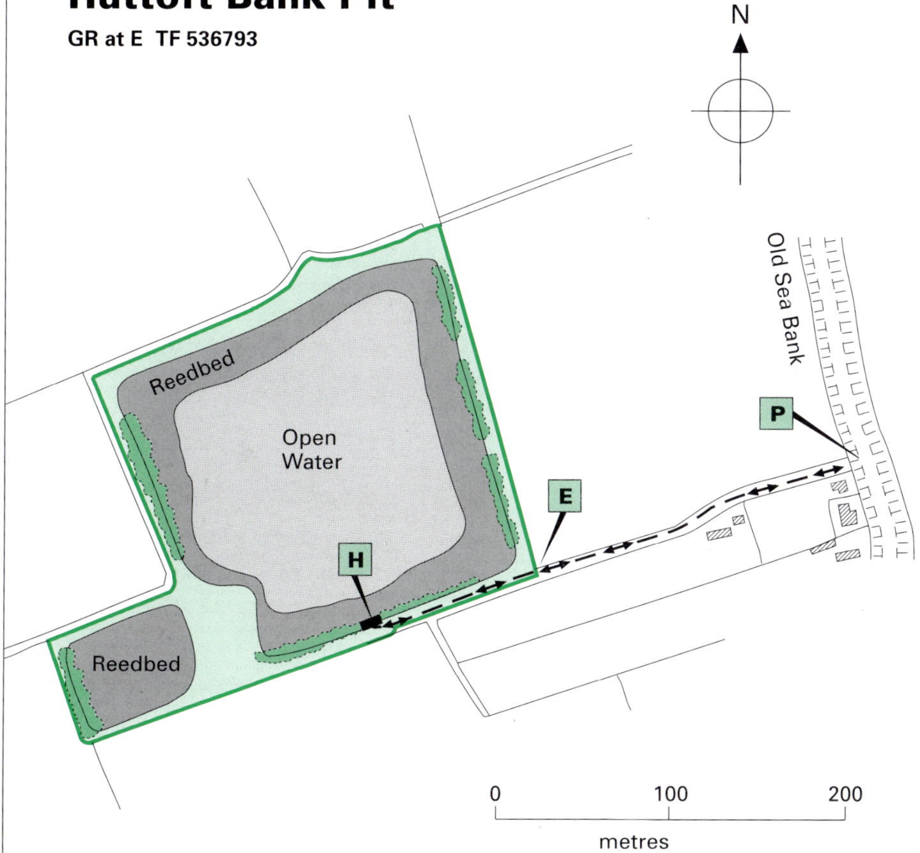

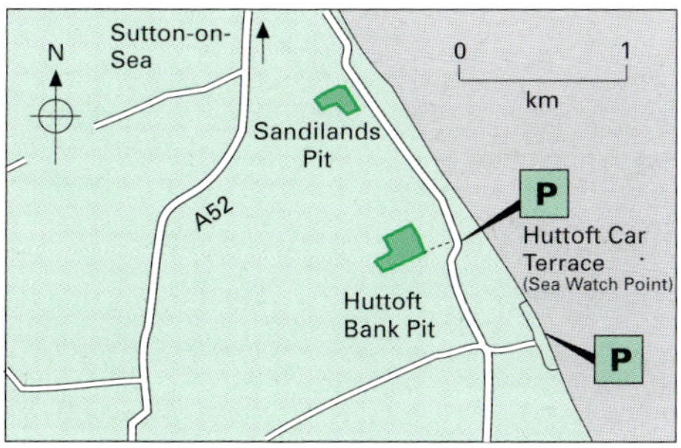

Keal Carr

East Keal and Toynton All Saints
OS: 122 GR: TF 385648 11.9 hectares (29.4 acres)
Part management agreement, part freehold 1980, part gift from Mrs DF Holmes
Map ref: 65

Location and Access

Situated to the north of the A16 road between Spilsby and East Keal, the reserve is reached by way of a narrow road (signposted to Old Bolingbroke and Mavis Enderby) that turns off the A16 a short distance east of East Keal. There is room for parking on the right-hand verge after a sharp left bend close to a Trust notice. Walk 300 yards down the track, turning right through a gate and crossing the field to the woodland edge. It is best to walk along the field edge to the corner where a hand gate gives entry to the wood. Visitors are requested to keep to the waymarked route. There is no circular route: it is necessary to retrace your steps. It is not a suitable reserve for visits by large numbers or anyone with disability.

Description and Management

This is probably the finest surviving example of the alder carrs of the southern Wolds, growing in small valleys created by streams cutting down through the soft Spilsby Sandstone into the Kimmeridge Clay below. The feature is also represented in the Snipe Dales and Sow Dale reserves. The upper slopes on the sandstone have an abundance of bluebells and red campion, but below the spring line on the impervious clay the ground is permanently wet and the woodland consists chiefly of alder and willow, which have been traditionally coppiced. Marsh-marigolds carpet this woodland floor in spring. Moschatel is also abundant, and there are both native species of golden-saxifrage: the opposite-leaved and the much rarer and rather finer alternate-leaved. In more open places there is a tall herb community, with yellow iris, giant horsetail, meadowsweet and angelica. There is a rich and varied bird fauna.

Management aims are the restoration of woodland cover in one area of the valley, the restoration of a coppice cycle, and the removal of sycamore, which has become over-dominant in places.

Golden-saxifrage: opposite-leaved (above) and alternate-leaved (below)

Keal Carr

GR at E TF 385648

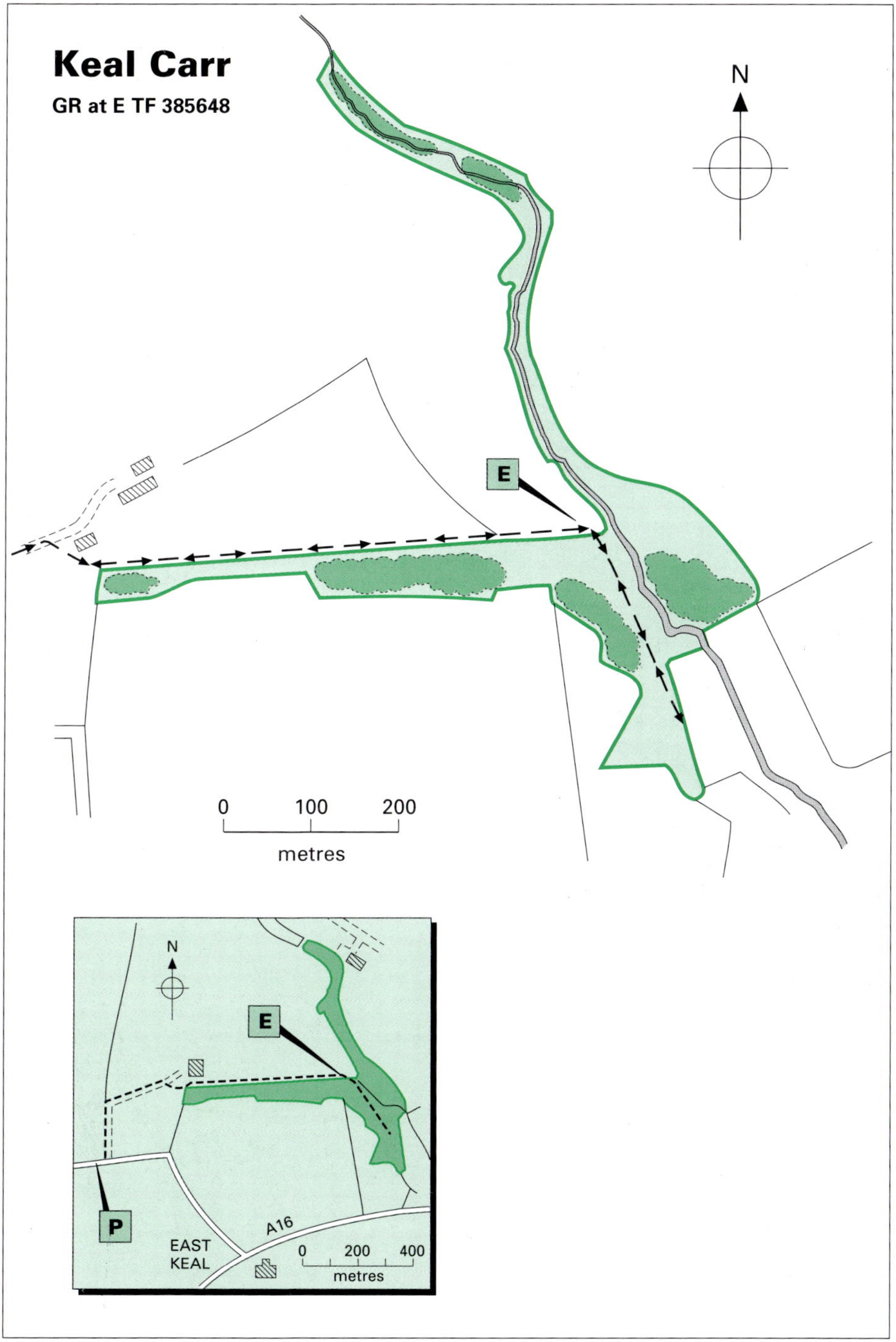

Killingholme Haven Pits

North Killingholme
OS: 113 GR: TA 165199 32 hectares (79 acres)
Management agreement 1979
Map ref: 27

Location and Access

The reserve lies 5 km (3 miles) north-west of Immingham Dock on the Humber Bank. It is situated to the south of Haven Road on the approach to North Killingholme Haven. There is no general access to the reserve, but adequate viewing points are available from the public road and bank-top footpath.

Description and Management

A complex of four flooded clay pits with associated rough grassland and scrub. Islands and promontories in the two larger lagoons, together with controlled water-levels, attract a variety and good numbers of waders and wildfowl. The reserve is important for its saline lagoon habitat, which supports several rare invertebrate species. The two smaller lagoons have deeper water and are fringed with reed and sedges; they provide suitable habitat for diving ducks, such as pochard, tufted duck and, occasionally, scaup. Breeding species include little grebe, and reed, sedge, willow and grasshopper warblers. The ruddy duck has also been known to breed here. It is the two large shallow pits, however, that are of the greatest importance for birds, particularly for migrant waders in spring and autumn. Spotted redshank, dunlin, greenshank, common sandpiper, little ringed plover, ruff and black-tailed godwit, the latter often in large numbers, are regular visitors. The list of scarce and rare species is long and includes spoonbill, avocet, little egret, little and Temminck's stints, red-necked phalarope, and curlew, pectoral, Baird's and white-rumped sandpipers.

Management is aimed at sustaining optimum conditions for waders and other species by maintaining shallow water and muddy margins.

Black-tailed godwit

Killingholme Haven Pits

GR at P TA 165199

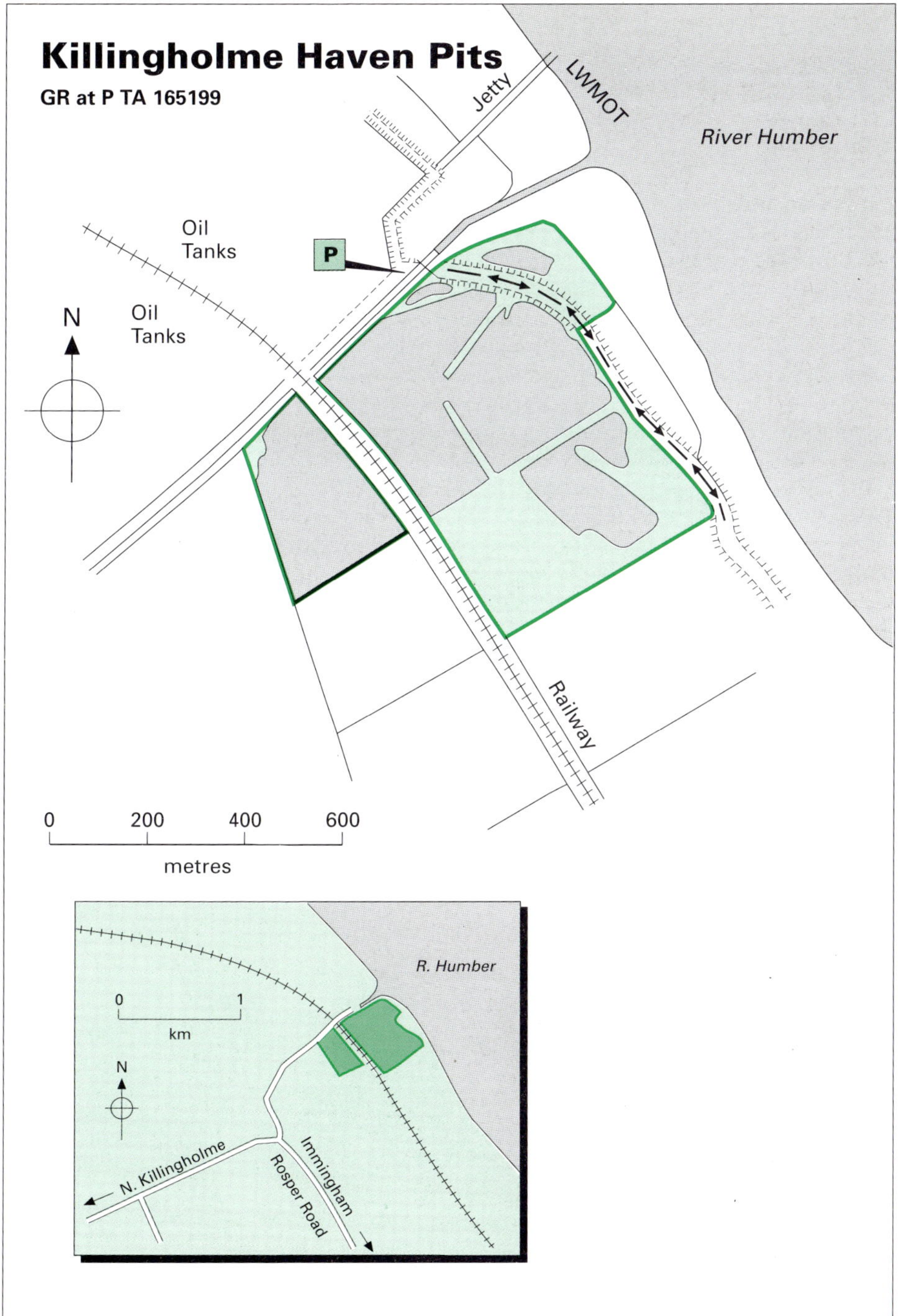

Kingerby Beck Meadows

Owersby
OS: 112 GR: TF 051941 9 hectares (22.3 acres)
Freehold 1989
Map ref: 88

SSSI

Location and Access

Lying between the Kingerby Beck and North Owersby near Market Rasen, the meadows are reached from the village of North Owersby by taking the turning in an easterly direction signposted to Glentham. A small parking area can be found on the left after about 1 km (0.6 miles). The entrance is via a stile or gate from the car park. Alternatively, the green lane adjoining the first meadow may be used to reach the lower meadows. Please take care to avoid trampling hay by keeping to the mown paths provided. Appropriate footwear is advised as it may be very wet along the line of the field drain crossing the centre of the meadow.

Description and Management

The four interconnected meadows slope gently towards the beck and are situated on sandy soils, which become heavier towards the beck. Kimmeridge Clay underlying the soils is near the surface and exposed in places along the banks of the beck. The composition of the flora varies from meadow to meadow, but there is a herb-rich sward, which includes bugle, fairy flax, yellow rattle, tufted vetch, cowslip, adder's-tongue, oxeye daisy, crosswort, pignut and many others. A particular speciality is a colony of frog orchids — a very rare species in Lincolnshire. Snipe may be seen in winter. Some of the boundary hedges are old and tall with species such as midland hawthorn, ash, crab apple and field rose.

The fields are cut for hay and the aftermath is grazed in the autumn.

Colony of frog orchids

Kingerby Beck Meadows

GR at E TF 051941

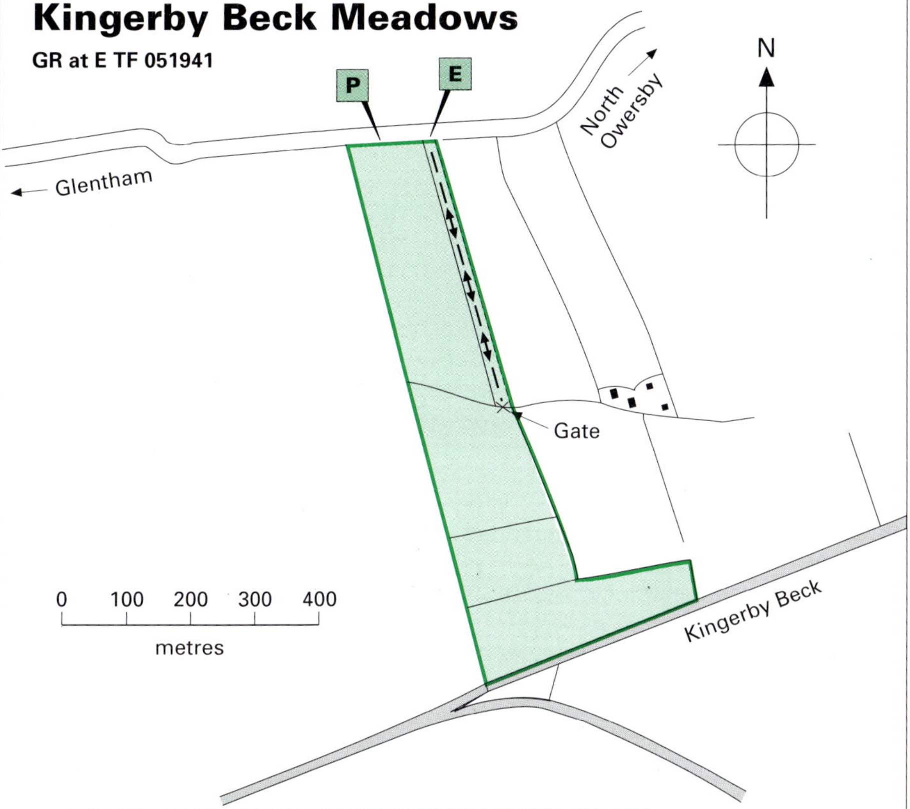

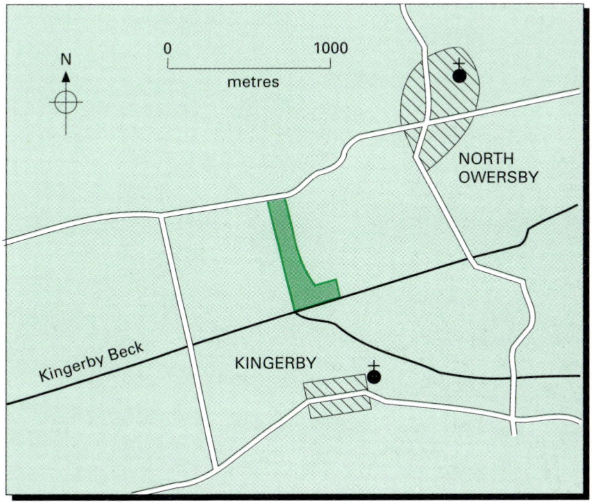

Kirkby Gravel Pit

Tattershall Thorpe
OS: 122   GR: TF 236609   14.8 hectares (36.5 acres)
Freehold 1991
Map ref: 92

Location and Access

Situated on both sides of the River Bain about 1.5 km (0.9 miles) south of Kirkby-on-Bain, the reserve lies to the east of a minor road that connects Kirkby-on-Bain with Tattershall. In order to minimise disturbance there is no access to the eastern shore, but the whole reserve may be viewed from the birdwatching hide situated in the willows towards the southern end. A small car park is provided.

Description and Management

Much of the reserve consists of worked-out shallow sandpits with some grassland, willow scrub and recently planted trees. It is divided into two unequal parts by the River Bain, which flows from north to south, and it is bounded on the west by the road and on the east by the Horncastle Canal. The reserve is an excellent bird-watching site. In spring and summer the shallow water, with its numerous spits and islands, suits many breeding birds, such as oystercatcher, ringed and little ringed plovers and common tern. Lesser black-backed gulls have also been known to nest here. At migration times many waders occur. Dunlin, greenshank, spotted redshank, green and wood sandpipers and ruff are regular visitors, and Temminck's stint has been recorded on several occasions. In winter it is a refuge for mallard, feral greylag geese, tufted duck, shoveler and many other ducks. Kingfishers and yellow wagtails nest in the area, and there is a thriving sand martin colony.

Management is designed to maintain shallow water and to limit the encroachment of scrub through control of water-levels and by rotational cutting of willows and other marginal vegetation.

Little ringed plover

Kirkby Gravel Pit

GR at E TF 236609

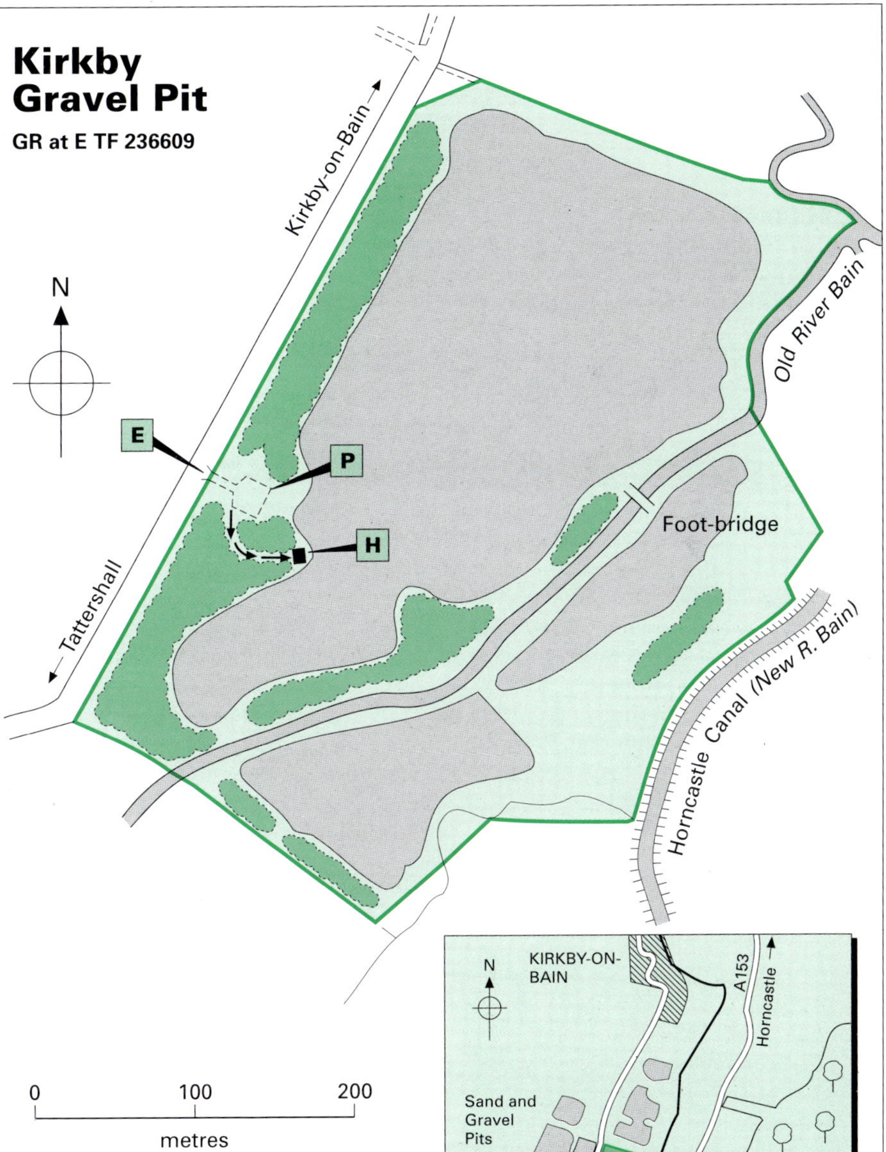

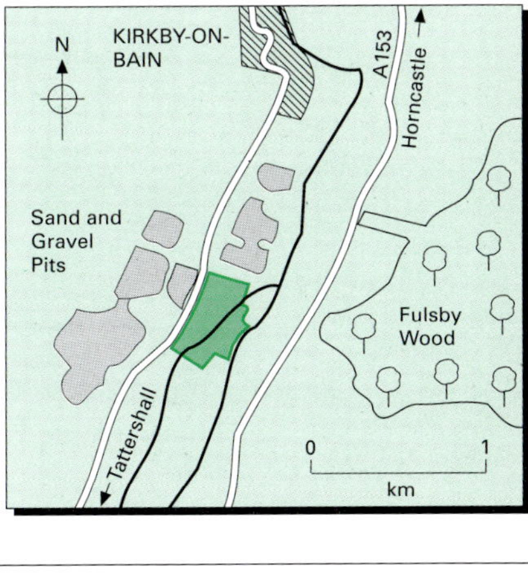

Kirkby Moor

Kirkby-on-Bain and Tattershall Thorpe
OS: 122 GR: TF 225629 75 hectares (185.2 acres)
Freehold 1969
Map ref: 28

Location and Access

Situated in the parish of Kirkby-on-Bain, the reserve lies on both sides of Moor Lane, which runs from Kirkby-on-Bain to Woodhall Spa. The entrance gate to the main part of the reserve is opposite Wellsyke Lane, about 1.5 km (0.9 miles) from Kirkby-on-Bain village. Cars can be parked either on the road verge outside the reserve or in the car park at the end of the private road through the reserve, but the road is unsurfaced and needs care. Visitors are requested to keep the gate closed at all times and to keep to the waymarked route or other defined footpaths. Adders occur on the reserve.

Description and Management

The reserve is the largest remnant of the once extensive heathlands of the Woodhall district. The sands and gravels here were deposited in delta conditions on the edge of the great Fenland Lake during the last Ice Age. As a result of leaching of mineral nutrients podsol soils have developed.

A large part of the reserve is dominated by heather, bracken and wavy hair-grass, together with moss and lichen communities. The southern part of the reserve is low lying and contains a lake originally made as a reservoir. There is oak, birch and pine woodland, and developing birch, hawthorn and sallow scrub with rowan and alder buckthorn. This diversity of habitat produces a wide range of plants and animals: 250 plant species have so far been recorded, together with 275 moths, 20 butterflies and 11 dragonflies. About 60 species of birds breed on the reserve in most years and over 100 species have so far been recorded.

The main aims of management are to: maintain the heathland and encourage more heather and wet heath conditions; reduce the area of bracken; control the spread of scrub onto the heath, while maintaining some developing scrub; and sustain the floristic diversity of some areas by regular mowing and by grazing with sheep.

Waymarked Route

Short route: 1.2 km (0.7 miles); long route: 2 km (1.2 miles)

The waymarked route starts from the car park, which is situated at the end of the trackway through the reserve, close to Foxhill Cottage. Cross over the ditch via a concrete footbridge. The ditch is well worth a closer look: a haunt of dragonflies and damselflies, it is covered in spring and summer by the floating leaves of the broad-leaved pondweed. Whirligig beetles and pond skaters swarm on the surface, and aquatic life abounds beneath.

Follow the waymarked route until you come to a diversion on the left, which leads to a hide overlooking the mere. Little grebe, coot, moorhen, mallard, and sedge and reed warblers regularly breed on and around the mere, and numerous

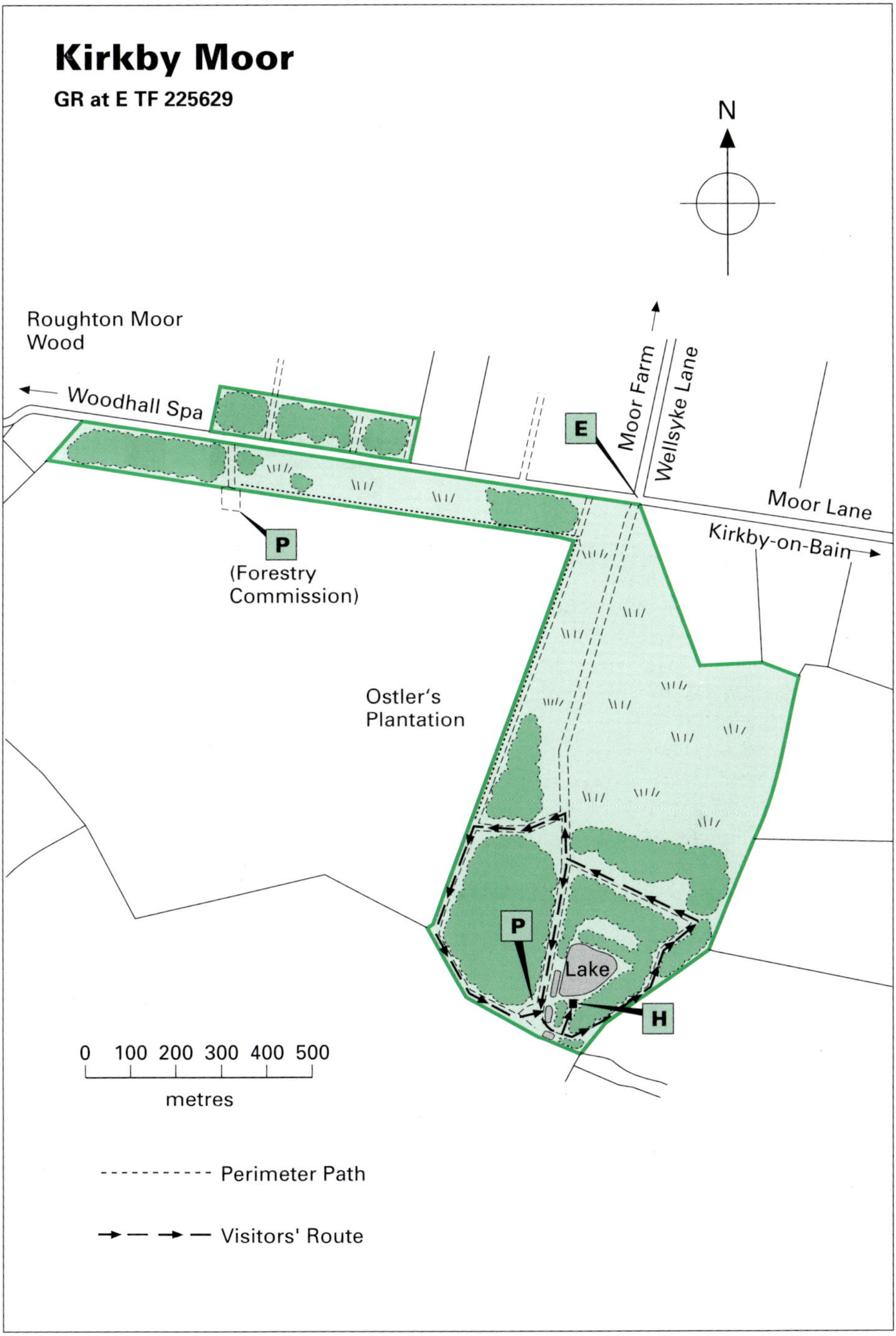

Kirkby Moor
GR at E TF 225629
N
Roughton Moor Wood
Woodhall Spa
Moor Farm
Wellsyke Lane
E
Moor Lane
Kirkby-on-Bain
P
(Forestry Commission)
Ostler's Plantation
P
Lake
H
0 100 200 300 400 500
metres
Perimeter Path
Visitors' Route

other birds can be seen, including kingfisher.

On leaving the hide retrace your steps to the ride through the wood. In spring or summer nightingale, chiffchaff, willow warbler and blackcap in the woodland behind the hide can be heard, if not seen. Jay, and blue and great tits can be seen in this area throughout the year. The route then follows the reserve boundary for some way alongside mature pines frequented by grey squirrels and goldcrests. It then turns left through an area of heather and winds through birch and sallow scrub, where nightingales sometimes nest. The birch and sallow give way to hawthorn, where common and lesser whitethroats, yellowhammer and turtle dove can be seen and heard.

The route then turns left through a marshy area, which is often waterlogged in winter. This area is at its best in summer when marsh-orchid, cross-leaved heath, marsh thistle, marsh valerian and devil's-bit scabious are in bloom, and it is probably the best place on the reserve for butterflies and damselflies, which are always to be found here on sunny days in their seasons. The rare bog bush-cricket can often be seen sitting on the seed heads of the common rush, and five species of grasshopper occur in this area.

The path then meets the road. For the short route, turn left along the road to return to the car park. For the long route, turn right, crossing a stile beside the cattle grid. This northern heathland area of the reserve is enclosed by fencing and is grazed by the Trust's flock of Hebridean sheep. After a short distance turn left along a track through another area of heather, with common centaury and stork's-bill. On the left are the shallow wells that once produced the district's water supply. Kestrels can often be seen hovering over the open heath. This is a good place to see woodcock roding in the spring and green woodpeckers at any time of the year. Wood sage occurs quite commonly; its greenish-yellow flowers are attractive to bumblebees: seven species seen feeding on these flowers have been recorded.

On reaching the reserve boundary the route turns left along a woodland ride. The birch woodland is rich in fungi, and in some years there are magnificent specimens of the attractive but poisonous fly agaric. Willow tits frequent the birches, and in the winter redpolls and siskins can sometimes be seen. This path can be very wet in the winter months and therefore appropriate footwear is advisable. The route then skirts a new plantation, where great spotted woodpecker, treecreeper, wood lark, tree pipit, warblers and spotted flycatcher are often to be seen in the fringing trees. Adders like to sun themselves by this path but in order to see them it is necessary to walk very quietly and carefully. The path then crosses the stream and returns to the car park.

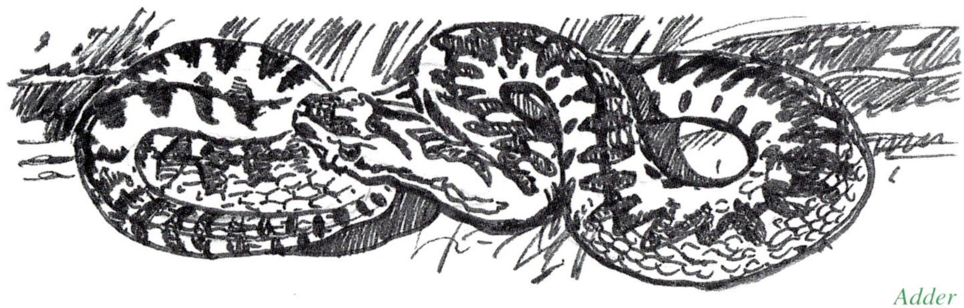

Adder

Kirkby-la-Thorpe Pit

Kirkby-la-Thorpe
OS: 130 GR: TF 082453 .1 hectare (2.5 acres)
Management agreement 1989
Map ref: 84

S

Location and Access

This small railway borrow-pit near Sleaford lies at the junction of the Sleaford–Boston and Sleaford–Spalding railway lines. Due to the difficulty of access and the terms of the Trust's agreement to manage railway land, access is by special permit only.

Description and Management

This flooded borrow-pit has become largely colonised by fen vegetation, dominated by lesser reedmace and blunt-flowered rush. There are also areas of tussocky grassland and willow scrub, supporting twayblade, common spotted-orchid, ragged-robin and adder's-tongue.

Management consists of clearing invading scrub from the fen, mowing parts of the grassland and cutting patches of fen vegetation.

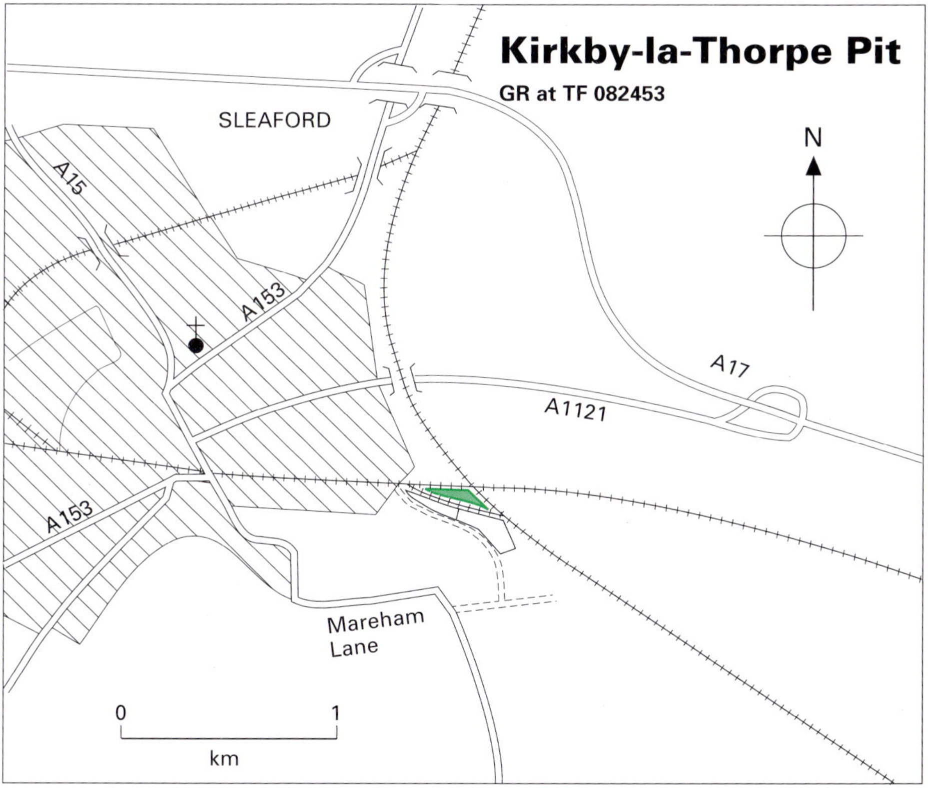

Langholme Wood

Haxey
OS: 112 GR: SK 751978 11.3 hectares (27.9 acres)
Freehold 1958
Map ref: 29

Location and Access

The wood is situated to the south-west of Haxey, on the west side of Langholme Lane. Turn west off the Misterton–Haxey road (A161) at the Nottinghamshire/North Lincolnshire border. After 1 km (0.6 miles) turn right onto a track that runs along the eastern edge of the reserve. There is space for cars near the main gate at the north-east corner. Visitors are requested to keep to the waymarked route.

Description and Management

Langholme Wood is a birch/oak woodland on Coversands with open heathland areas. The woodland has developed on abandoned farmland and heath during the past 60–70 years. It shows succession from open heathland to birch scrub and birch/oak woodland. The understorey is mainly elder and hawthorn with bramble and stands of gorse and broom. There is a ground cover of heather and wavy hair-grass in the open areas. A good variety of fungi can be found. The wood attracts many birds; 52 species have been recorded. Management of the reserve is aimed at maintaining the open heathland and improving the woodland by selective clearing and thinning of birch to increase woodland edge habitats.

Waymarked Route 1.3 km (0.8 miles)

Enter the reserve via the stile near the main gate and follow the waymarked route past gorse and bramble scrub. Here you may see wren, robin, long-tailed, great, blue and coal tits and a variety of finches. After about 100 m you will reach a hut. You may see redpolls here and, in the summer months, spotted flycatcher. Follow the route in a westerly direction to a glade. Then proceed through young silver birch and enter an area of open heathland with heather and various mosses. Leaving the open area, follow the path through the birch and oak trees. Look out for green woodpecker, great spotted woodpecker, jay and treecreeper.

Continue on the path across the western edge of the reserve. Turn left along the edge of the plantation. In 1983/84 the sycamore trees in this area were cleared and replanted with native trees — mainly oak, ash, beech, gean, rowan, Scots pine and field maple. On reaching the eastern end of the plantation after about 50 m turn right and there is another area of young trees to your left, which attracts many small birds. In the summer warblers, such as willow warbler, chiffchaff, whitethroat, blackcap and garden warbler, can be seen or heard. Follow the path at the eastern edge of the plantation until the path turns sharply left. After 200 m the path turns sharply right between two large oaks. Follow the path for about 200 m and then turn left where you will see coppiced willow to your left. The path follows this southern part of the reserve through willow and birch. After passing an area of bracken on the right the original path will be joined some 20 m from the reserve entrance.

Langholme Wood

GR at E SK 751978

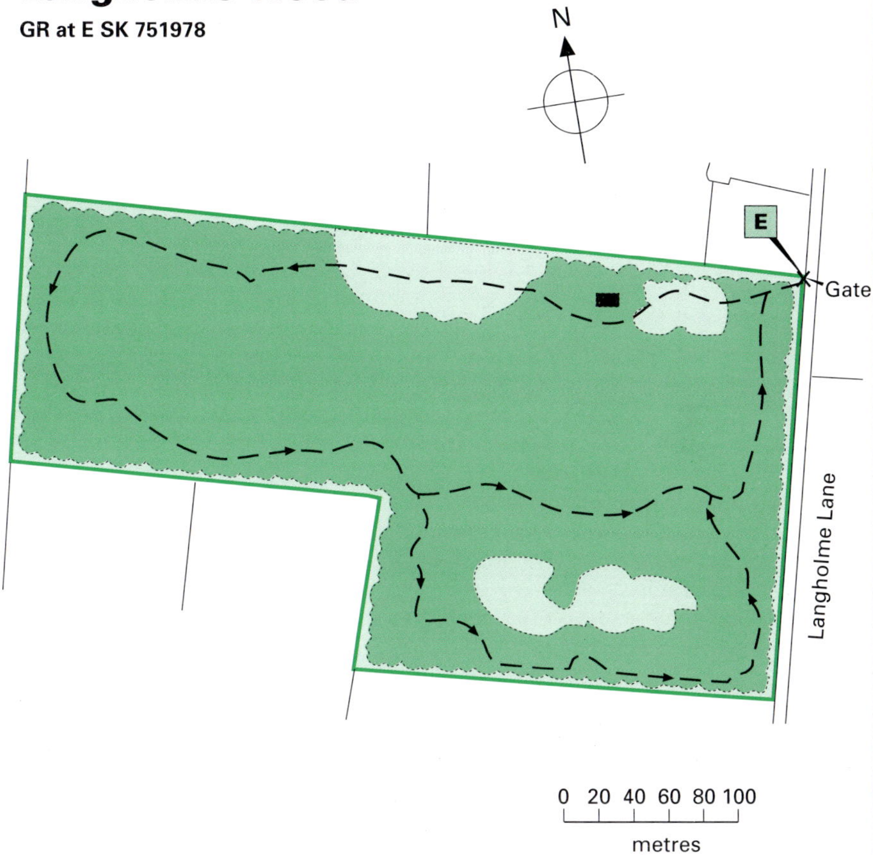

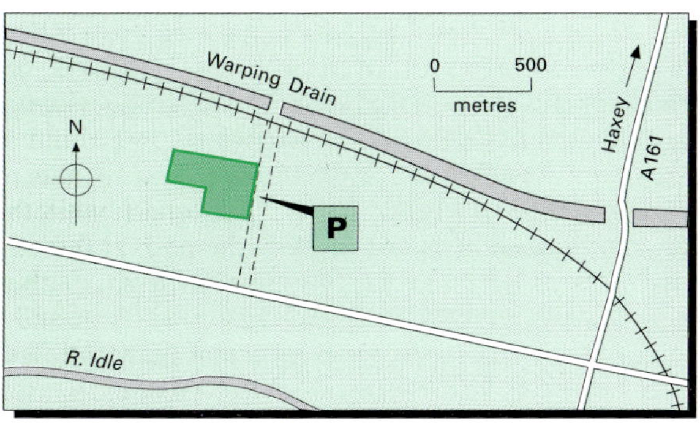

Lawn Wood and Meadows

Castle Bytham
OS: 130   GR: SK 993193   12.2 hectares (30.2 acres)
The Harris Memorial Reserve. Freehold 1993
Map ref: 102

The meadows were given to the Trust in 1993 by Mrs Mary Harris as a memorial to her husband, David Harris. Lawn Wood was purchased in 1995.

Location and Access

The reserve is situated north-east of Castle Bytham and is reached from an unclassified road (Counthorpe Lane) linking Castle Bytham and Creeton. Limited parking is available on the verge near a bend in the road. Please do not obstruct the farm track with vehicles. Follow the track on foot about 250 m to the entrance. The entrance (GR: SK 996193) is through a gate or stile.

Description and Management

The reserve consists of two meadows, named Bottleneck and Jackson's paddock, and the adjoining Lawn Wood.

The name Bottleneck aptly describes the shape of the first field, which lies to the north-west of Lawn Wood. Jackson's paddock (which derives its name from a former owner) links up with Bottleneck and overlooks the village of Castle Bytham with its ancient castle site. Both fields are on heavy clay loam and are wet in places. Tall hedges with field maple, hawthorn, blackthorn and rose serve as a colourful surround and shelter both fields, providing a haven for ringlets and meadow brown butterflies in the summer. The summer also brings a magnificent display of buttercups, with those in the shaded parts of Bottleneck showing a slightly later flowering period. Yellow rattle and many other meadow species are present.

Lawn Wood is an ancient deciduous wood consisting mainly of oak and ash, with field maple, midland hawthorn and the scarce wild service tree. The ground flora includes species of old woodland, such as wood anemone, woodruff and early-purple orchid. Fallow and red deer are frequently to be seen in the wood and meadows.

The meadows are cut for hay in July, and the aftermath is grazed by sheep. In the wood, thinning operations are designed to restore a varied ground flora. Parts are coppiced, while the ride system is being extended to provide edge habitat for birds and butterflies. The boundary between the wood and the meadows provides particularly important habitat, which is being carefully looked after in order to retain and extend the scrubby margins.

Lawn Wood and Meadows

GR at E SK 993193

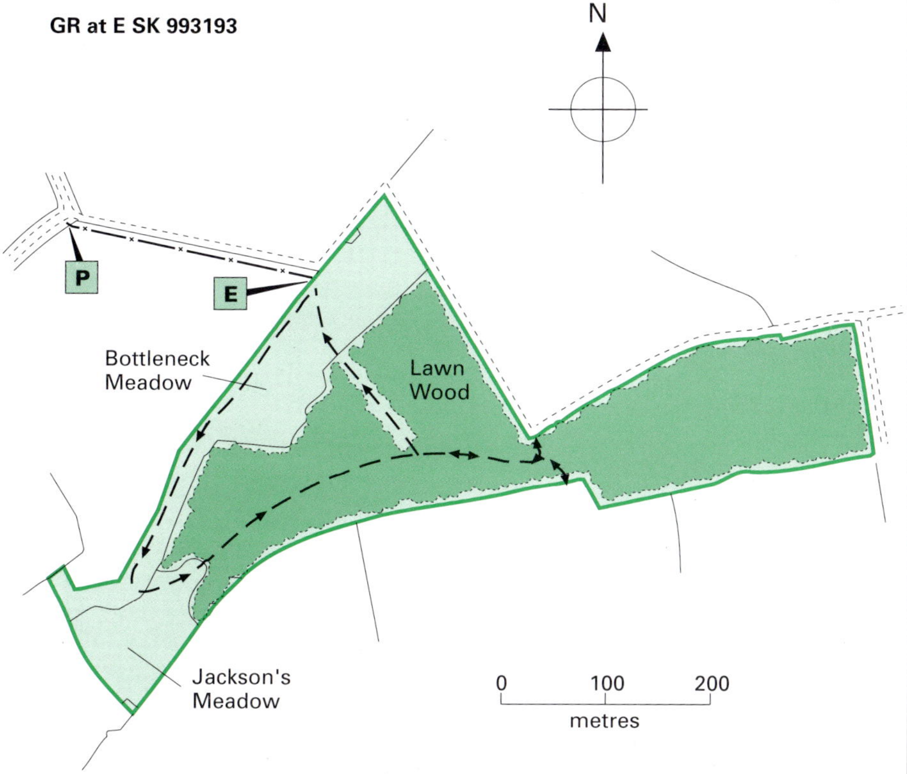

Permissive Footpath (Lincs C.C.)

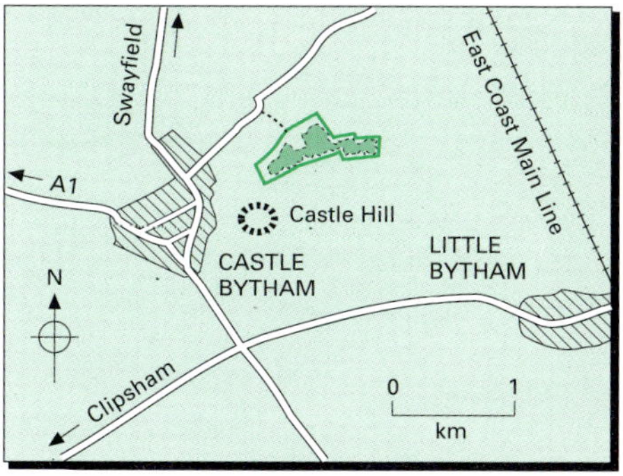

Lilley's Carr

Toynton All Saints
OS: 122 GR: TF 387635 2.6 hectares (6.4 acres)
The Mark Newlands Lilley Memorial Reserve. Leasehold 1994
Map ref: 95

Location and Access

The reserve is situated on the extreme southern slope of the Wolds in the parish of Toynton All Saints. Turn off the A16 about 1.5 km (0.9 miles) south-west of Spilsby onto the road signposted 'Toynton All Saints/Toynton St Peter'. After about 1 km (0.6 miles) turn right into Watermill Lane (opposite the church) from where the reserve can be accessed. There is parking space at the entrance to the old sand pit on the left, after 250 m.

Follow the footpath to East Keal for 30 m and then turn left at the gate to a former quarry. Continue along the field boundary footpath for about 300 m to the reserve entrance. Access to the north of the entrance is via a stile at the side of the gate, which leads into a grassland enclosure. The long strip bordering the stream to the south is boggy and has high vegetation and is therefore best seen by following the field boundary path.

Description and Management

The reserve contains the last of the series of alder carrs of the southern Wolds before the start of the Fenlands. It is on the same stream as the Keal Carr reserve, which is about 1 km (0.6 miles) away, with another carr between the two reserves. The reserve is therefore of importance in maintaining the system of carrs and their associated habitats.

The reserve also contains about 1.2 hectares (3 acres) of rough grassland on the Spilsby Sandstone and Kimmeridge Clay. The grassland was formerly ploughed and drainage has been attempted. It therefore lacks floristic variety, and is being invaded by rushes, gorse and an extension of the alder carr. The introduction of appropriate grassland species is planned as part of reserve management. The tongue of land extending south from the grassland alongside the stream is dominated in summer by a dense growth of butterbur, and is almost impassable except in winter and spring.

Nesting birds include blackcap, garden warbler, whitethroat, blue, great and willow tits, spotted flycatcher and treecreeper, while sparrowhawk and hobby are occasionally seen.

Management mainly involves limitation of scrub extension and thinning of over-dense alder. The grassland enclosure is grazed by sheep during the summer months.

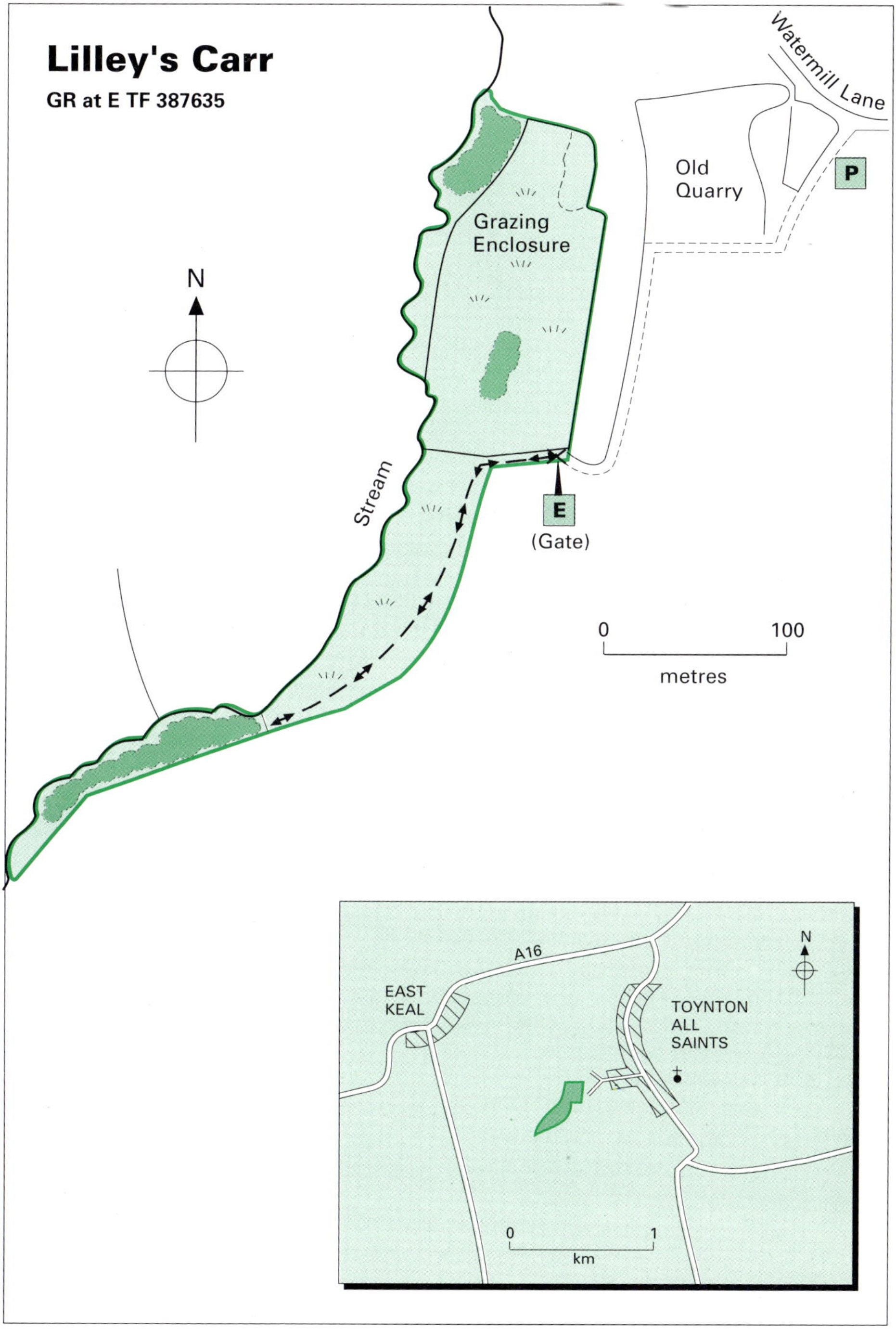
Lilley's Carr
GR at E TF 387635
Watermill Lane
Old Quarry
P
Grazing Enclosure
N
Stream
E
(Gate)
0
100
metres
A16
EAST KEAL
TOYNTON ALL SAINTS
N
0
1
km

Linwood Warren

Linwood
OS: 121 GR: TF 133877 27.5 hectares (68 acres)
The RK Cornwallis Memorial Reserve. Freehold 1957
Map ref: 30

SSSI

Location and Access

The reserve is located about 2.5 km (1.5 miles) east of Market Rasen on the south side of the Legsby road opposite the golf course. Cars may be parked at the entrance or on the roadside. Linwood Warren is dedicated to the memory of the late Dick Cornwallis, Chairman of the Trust from 1962 to 1968, and nearby landowner and farmer.

Description and Management

The reserve forms part of a once more extensive area of heathland, the remainder of which is now largely afforested or occupied by the golf course. The Warren has been put to a variety of human uses in the past, evidence of which is provided by the presence of flint implements, Romano-British pottery kilns, banks of a medieval or later field system, the butts of a 19th century rifle range, and 1914/18 practice trenches. The result of these various human activities is a mosaic of habitats that gives the reserve a varied flora and fauna. The high ground to the south is clay and carried mature oak woodland (with some small-leaved lime or linden tree, from which the name Linwood is derived) until it was felled more than 40 years ago. It is now a Forestry Commission plantation largely of pine. To the south-east, one area of pine has been felled, leaving natural regeneration.

The roadside strip of oak woodland extending from the reserve eastwards to the corner at Little London Wood is managed by the Trust by agreement with the Forestry Commission.

The flora and fauna of the reserve are rich and varied. All three heathers (bell heather, cross-leaved heath and ling) occur, and there are damp areas rich in mosses and lichens — including the Iceland moss, a rare northern upland species which is a relict here on a lowland heath. Twenty-one species of butterfly and well over 200 moths have been recorded. Water shrew and badger occur, as well as adder, grass snake, slow worm and common lizard. Of the 70 or so species of birds recorded over 30 are known to have bred. Notable among these are woodcock, tawny owl, green and great spotted woodpeckers, sparrowhawk, tree pipit, redstart and grasshopper warbler. Crossbill, snipe and lesser spotted woodpecker have also been seen on the reserve.

Waymarked Route

From the entrance gate follow the track past the RK Cornwallis memorial. On the left is a large patch of bilberry — a common plant of upland moors, but unusual on a lowland heath. The path emerges onto open heathland, where the left fork follows a circular route through the eastern part of the reserve and the right fork loops through the western part.

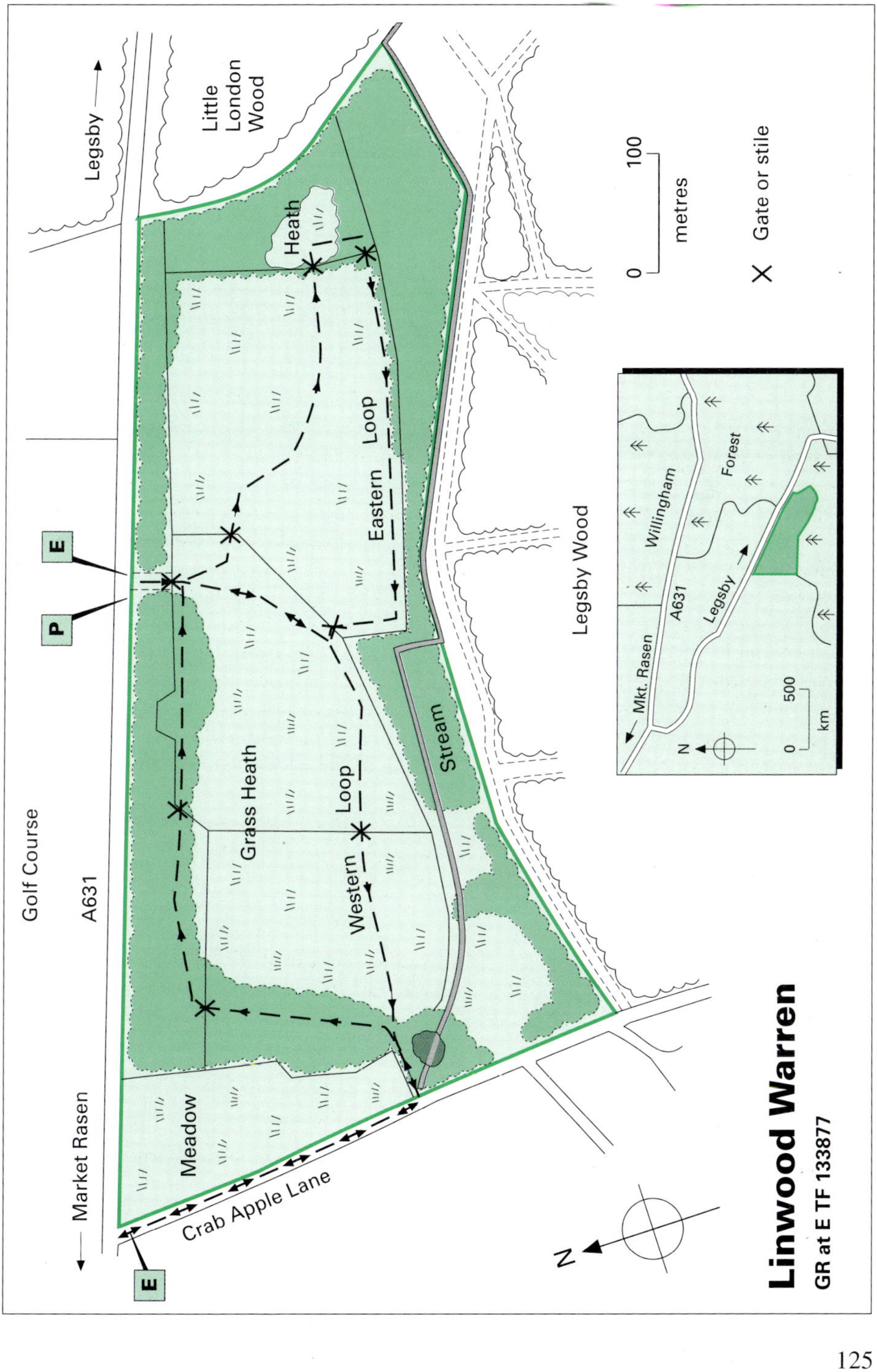
Linwood Warren
GR at E TF 133877
Legsby
Little London Wood
Heath
Golf Course
A631
E
P
Eastern Loop
Grass Heath
Western Loop
Stream
Legsby Wood
Meadow
Market Rasen
Crab Apple Lane
E
N
0
100
metres
Gate or stile
Willingham
Forest
A631
Legsby
Mkt. Rasen
N
0
500
km

Western loop: 1 km (0.6 miles)

The path crosses open grassland dominated by wavy hair-grass. The scattered pines and backdrop of woodland provide excellent habitat for green woodpecker and tree pipit. Keep straight ahead and enter the grazing enclosure by the stile. Grazing here has been extended onto the heath in order to create more diversity. The bare areas are experimental plots for re-establishing heather. Cross the enclosure to another stile. From here there is a choice of routes. To return via the road, taking in the small pond and the meadow at the western end of the reserve, cross the stile. The pond attracts dragonflies and damselflies, and water vole may be seen. The meadow is wet and supports species such as marsh violet, sneezewort and devil's-bit scabious. Alternatively, turn right and follow the path through the wood where there are some fine Scots pine planted at the beginning of the century. Look out for willow and coal tits, goldcrest and great spotted woodpecker. The path returns to the entrance track close to the Cornwallis memorial.

Eastern loop: 1.2 km (0.75 miles)

From the entrance take the left track after passing the Cornwallis memorial. The broad track leads through pine and birch, passing a hut on the right. The area by the hut has now been cleared and heather seedlings are re-appearing. The path emerges on the eastern heath, where cross-leaved heath and bell heather grow, together with cottongrass, creeping willow and many fungi, mosses and lichens. Please avoid excessive trampling on the lichen heath. The Trust has greatly extended the lichen heath by clearing encroaching birch and pine. The heath is grazed by the Trust's Hebridean sheep from May to September to control invasive birch scrub. The path crosses the heath and enters the woodland where it turns right to meet an old boundary bank. Follow the bank to the right through the oak and birch woodland. Honeysuckle and lily-of-the-valley are present here. Look out for goldcrest and great spotted woodpecker and, in summer, garden warbler. The bank comes to an end, but the path continues for a short way before turning right along a track, which is the route of an ancient roadway. The path soon turns right again, crossing an open tract of wavy hair-grass to return to the entrance track.

(Left to right) Ling, bell heather and cross-leaved heath

Little Scrubbs Meadow

Langton-by-Wragby
OS: 121 GR: TF 145744 2 hectares (5 acres)
Management agreement 1972
Map ref: 31

Location and Access

The meadow is situated within Bardney Forest near Wragby on the eastern side of Little Scrubbs Wood (part of the Forestry Commission's Chamber's Wood complex). Cars should be left in the Forest Nature Reserve car park at Chamber's Farm, which is signposted from Wragby. From the car park, join the forest walks footpath and follow the white markers to Little Scrubbs Meadow.

There are two other Forestry Commission waymarked forest walks starting from the car park. There is a butterfly garden and a Visitor Centre, which is open some weekends in the summer whenever volunteers are available. A special trail guide is available. Most of the Forestry Commission woodlands in this part of Bardney Forest are ancient limewoods of great national importance, and are managed by the Commission as a Forest Nature Reserve. See also the entry for Hatton Meadows on page 92.

Description and Management

The reserve consists of old grassland on heavy clay soil of a type now rare in the county. Meadow plants, such as great burnet, saw-wort and devil's-bit scabious, are abundant, while the green-winged orchid survives in small numbers. A woodland edge on all sides enhances the value of the reserve for birds and insects. Woodcock, great spotted and green woodpeckers, nightingale, grasshopper and other warblers nest in the adjoining woodland and scrub. In recent years 26 species of butterflies have been recorded, including purple hairstreak, white admiral and speckled wood.

Management consists of grazing with Hebridean sheep to control scrub, mowing in the autumn, and coppicing the woodland edge.

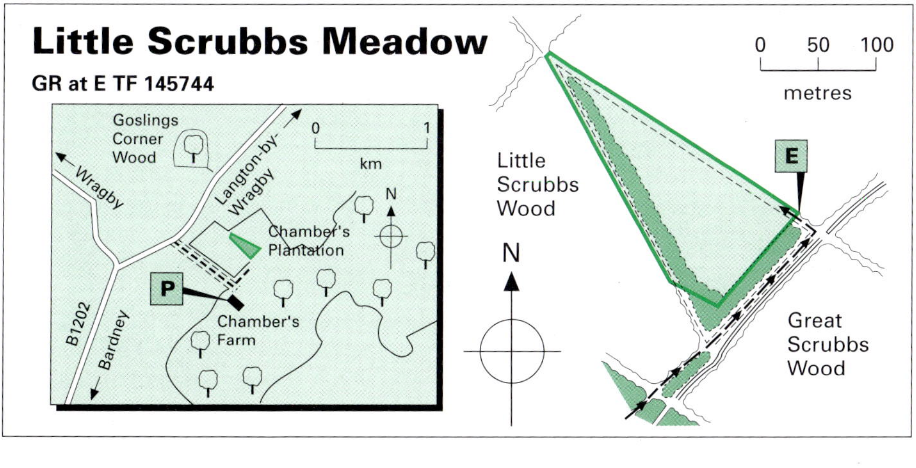

Messingham Sand Quarry

Messingham and Manton
OS: 112 GR: SE 908032 40.5 hectares (100 acres)
Leasehold 1981
Map ref: 62

Location and Access

The reserve lies to the east of the B1400 Messingham–Kirton road. The entrance is opposite Scallow Grove Farm. A track leads from the entrance to a small car park, where a waymarked route is indicated.

Description and Management

The reserve has been created by the excavation of sand, which has left a series of lagoons of differing depths with fringing beds of reed, reedmace and rushes. There are also remnants of the heathland from which sand was extracted and which supports heather and other heathland plants. Woodland and marsh add further habitat variety to the reserve, and this is reflected in the richness of wildlife, which includes 370 species of plants, more than 180 species of birds (73 of which breed), 20 species of butterflies, 16 species of dragonflies, over 285 species of moths and many other invertebrates. There are a number of rare and endangered species, including pillwort — a water fern — which occurs here probably in greater abundance than at any other site in Britain. Management consists of mowing the grassland in autumn.

Waymarked Route 1.2 km (0.75 miles)

From the car park follow the waymarked route, from which a great carpet of pillwort may be seen in the nearby lagoon. It then traverses an area of mixed grasses, dog daisies, vetches and bird's-foot trefoil. Many butterflies occur here, including wall, meadow brown, and large and small skippers, and also various damselflies.

As you reach the lake in front of you, take a small detour to the right to visit the grebe hide. Typical birds seen from here are coot, mute swan, ducks, geese and, of course, great crested grebe. Return back to the trail by the same route.

The path then traverses between the lake on the right and a conifer plantation with sallow scrub on the left, with common reed on both sides. Look for green-veined white, ringlet and meadow brown butterflies, and common blue and blue-tailed damselflies. Reed warblers and willow warblers will be heard if not seen.

As the path curves round between two lakes, look for reedmace and reed bunting. A little further on, there is a meadow to the right, which is surrounded by gorse, hawthorn, sallow, wild rose and bramble. The common blue butterfly may be seen here and also the black and yellow-striped caterpillars of the cinnabar moth on the ragwort.

Passing once again between the lakes, there are good views of wildfowl on both sides. The path then enters an area of hawthorn scrub where various butterflies, dragonflies and damselflies can be seen, as well as many birds, such as bullfinch, greenfinch, chaffinch, linnet and redpoll. Follow the path through a

Messingham Sand Quarry

GR at E SE 908032

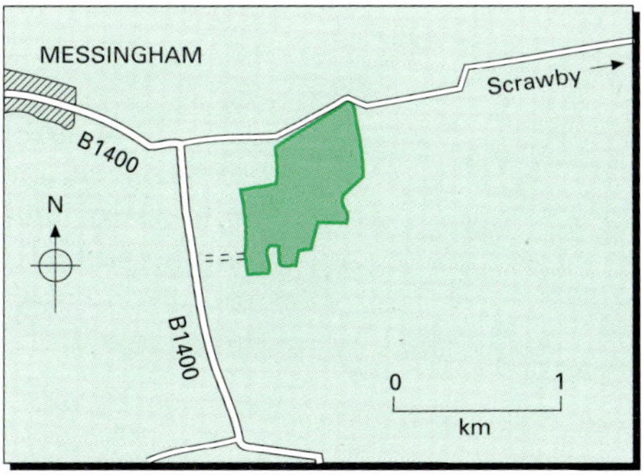

VP Viewing Point

small area of birch and oak woodland where great and blue tits may be seen. You will then enter the last remaining fragment of the original heathland. Typical heathland plants are heather, heath bedstraw, tormentil, sheep's sorrel, broom and gorse. Common lizards are often seen basking on the wooden seats, and small copper butterflies may be seen feeding on the heather.

The route continues between lakes flanked by gorse, sallow and some alder. Wall and peacock butterflies may be seen basking on the warm sand, while the gorse attracts dragonflies and damselflies. At the end of this stretch of path, turn right to visit the duck hide, which gives good views over the lake, marsh and surrounding countryside. To the left of the hide there is a breeding colony of black-headed gulls in the marsh. These waters are a haven for many wintering wildfowl, including teal, wigeon, mallard, pochard and, occasionally, Bewick's swan.

Returning back to the main path, the route once again leads you between gorse and sallows. If you turn right across the wooden bridge, a half-mile detour will bring you to the wader hide, where wading birds that may be seen include curlew, redshank, lapwing and ringed plovers. On the route to the hide, take the opportunity to look across the marsh at the view point, where you will be rewarded with a view of the orchids in June and July. Return back to the main route by the same way.

The route then skirts round the conifer plantation and meanders through the Scots pines. In the wetter areas, alder, willow and marsh plants grow. Common lizards may be seen on both the bridges that traverse ditches. Gatekeeper and ringlet butterflies can sometimes be seen. After leaving the wood, the route leads back to the car park.

Common lizard with small copper butterfly in foreground

Metheringham Delph

Metheringham and Blankney
OS: 121 GR: TF 133640 42.9 hectares (106 acres)
Management agreement 1988
Map ref: 86

Location and Access
The reserve runs from the Car Dyke near Blankney Wood towards the River Witham downstream of Southrey. Access and parking space is in the hamlet of Tanvats. There are several bridges and gravel fords allowing circular walks of between 2 and 8 km (1.2 and 5 miles), but the gravel crossings may become impassable at times of high water-levels.

Description and Management

The Delph is a 5.5 km (3.4 miles) length of relict channel that used to carry water across the Fens to the River Witham. As the water now drains into the Witham through other delphs, such as Timberland and Nocton, this channel has become redundant and, in the absence of dredging, has developed an abundance of wildlife. The delph is used as a fish nursery with roach, bream, perch and tench present along with the occasional pike. The aquatic vegetation is well developed with areas of reed, reed sweet-grass, bur-reed, reedmace, water dock, water mint, water forget-me-not and white water-lily. Birds include coot, moorhen, mute swan, heron, sedge and reed warblers and, occasionally, kingfisher. Barn owls, which breed in the area, hunt the banks. Tortoiseshell butterfly and the commoner brown butterflies are present, sometimes in large numbers. In winter many ducks, such as shoveler, teal, pochard and tufted duck, come to the Delph to feed.

Access along the Delph is maintained by mowing the taller vegetation as necessary.

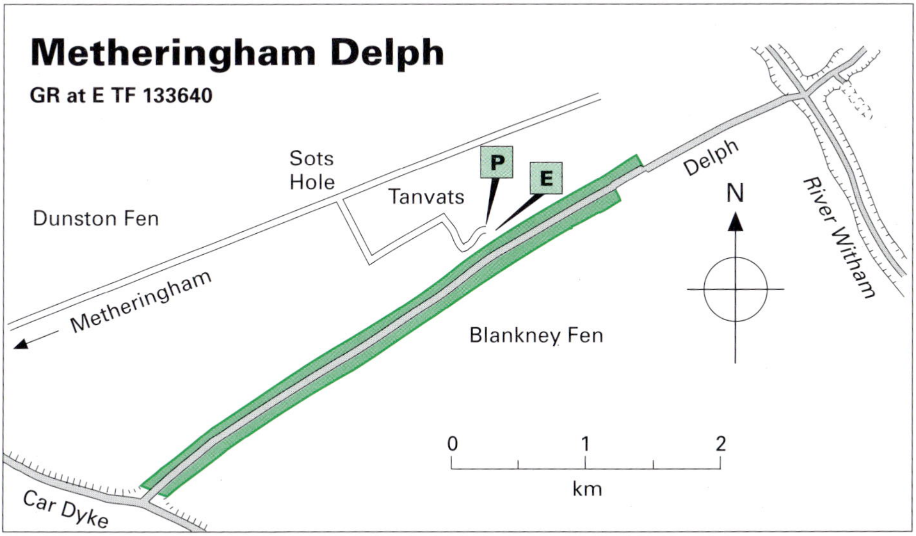

Mill Hill Quarry

Claxby-by-Alford
OS: 122 GR: TF 452716 2.4 hectares (6 acres)
Freehold 1968
Map ref: 33

Location and Access

The quarry lies on the north side of the Claxby–Willoughby road near the top of the hill to the east of Claxby village, and about 1 km (0.6 miles) from Hoplands Wood. Parking space is available at the entrance or on the roadside. The cliff face and scree should not be climbed, and care must be taken not to trample vegetation unnecessarily in summer. Beware also of exposed lime kilns. For a location plan with adjoining reserves see page 51.

Description and Management

Near to the eastern edge of the Wolds on a steep valley slope, the reserve consists of a disused chalk pit and a spinney. Already in use by the mid-18th century, the quarry produced stone for the marshland roads and, later, agricultural lime until work ceased in the late 1950s. Two old lime kilns survive, one of which is in particularly good condition. The cliff face at the north end shows the Middle and Lower Chalk with an intervening band of grey marl. The most abundant fossil is the bivalve mollusc *Inoceramus*, which has a strongly ribbed shell. The quarry exhibits various stages of colonisation by chalk vegetation: the screes and sparsely

Mill Hill Quarry

covered ground below the cliff; well established turf with a colourful variety of wildflowers, including purging flax, salad burnet, giant bellflower, burnet-saxifrage, cowslip and an abundance of marjoram, and pyramidal and spotted-orchids; and scrub and woodland of ash and beech. Breeding birds include spotted flycatcher, pied wagtail, tawny owl, four species of warblers and five species of finches. Sun-loving butterflies, such as wall, common blue and small copper, and many bumblebees frequent the warm, south-facing slope.

Management is designed to maintain the various types of habitats by mowing and scrub control, and the woodland has been diversified by introducing more beech. The reserve is of use for educational purposes as it has much that is of historical interest: the mill mound, which is the site of a long-vanished post-mill but originally a Bronze Age round barrow; the lime kilns; and the remains of a blacksmith's shop near the entrance. A lane on the east side of the quarry leads to the mill mound, which commands a splendid view over the marshland.

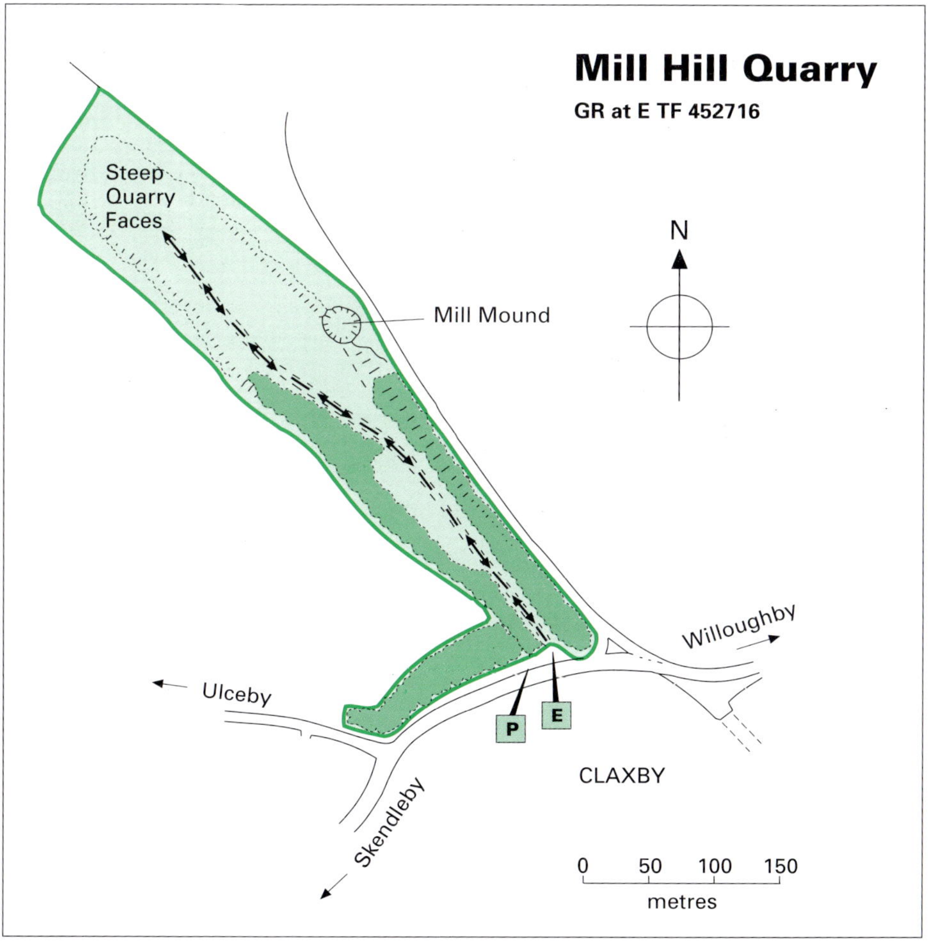

Moor Closes

Ancaster
OS: 130 GR: SK 982438 6.4 hectares (16 acres)
Freehold 1971
Map ref: 34

SSSI

Location and Access

The four fields of the reserve are close to the village of Ancaster. At the south end of Ancaster is the crossroads of the A153 Sleaford–Grantham road and Ermine Street. A short distance west of the crossroads, along the main road, a lane turns right and passes the west end of the church. The reserve entrance gate is reached on foot on the left after about 500 m. Parking space is available in the village near the church. The lane may be reached by walking through the churchyard. Visitors are requested to keep to the waymarked route. For a location plan with adjoining reserves see page 25.

Description and Management

These old pasture fields lie on the floor of the Ancaster Gap on sands and gravels carried from the Trent Valley through the gap in the limestone by glacial overflow waters. A small stream intersects the reserve, which is mostly low-lying and wet, although the land rises a little on the north side to give better drained and drier conditions.

The low areas have a rich, wet meadow and marsh flora with a variety of grasses, sedges and rushes and localised plants, such as marsh arrowgrass, ragged-

Ancaster Church from Moor Closes

Moor Closes

GR at E SK 982438

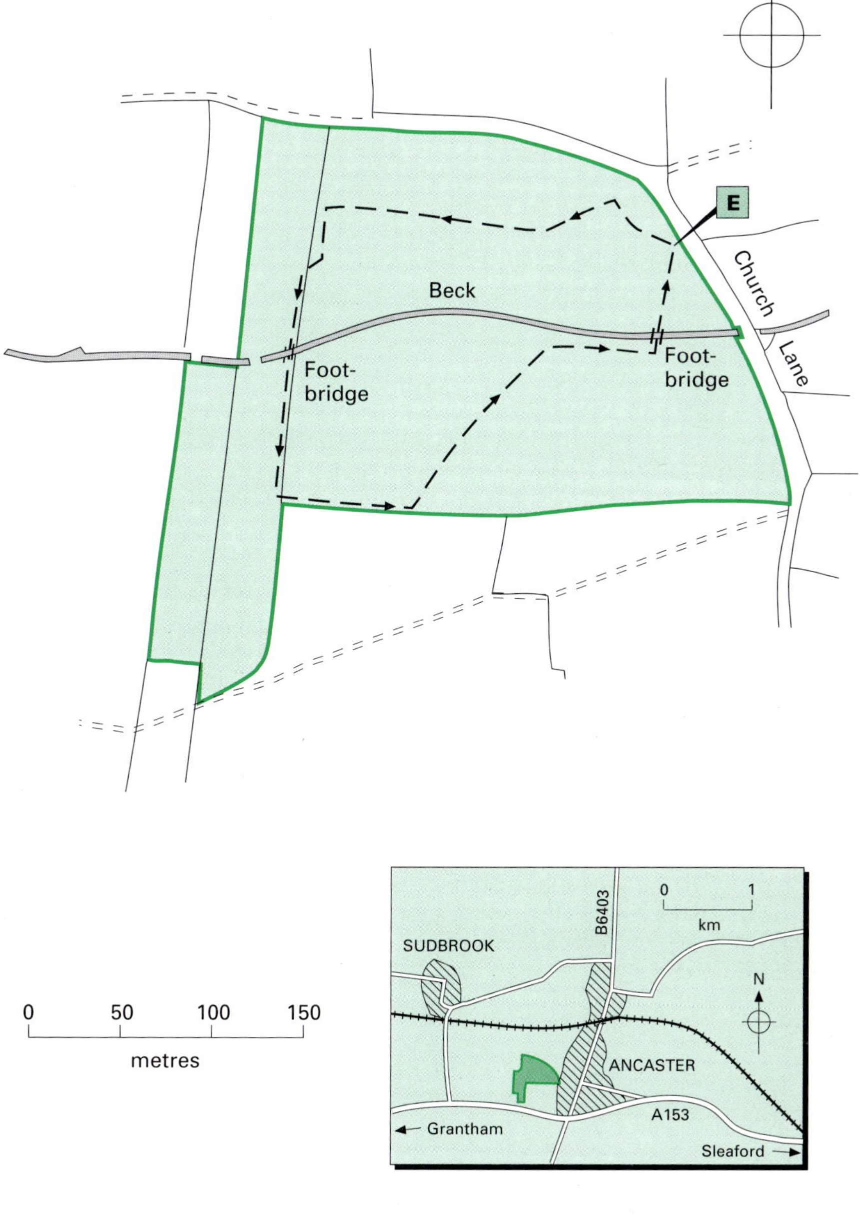

robin, marsh valerian, devil's-bit scabious, adder's-tongue, and an abundance of marsh- and spotted-orchids. The drier areas have a similarly varied flora, including dropwort, meadow saxifrage and the distinctive inland sub-species of thrift *Armeria maritima elongata*. This rare thrift, with long stems, formerly occurred on a number of sites on the old Lincoln Heath, extending into Leicestershire, but is now confined in Britain to the reserve and the nearby graveyard. The ecological interest of the reserve is further enhanced by the stream (where there are old pollarded willows) and by large hedges with a variety of trees and shrubs.

Traditional management by grazing with cattle is being continued in order to maintain the richness and interest of the site.

Waymarked Route 1 km (0.6 miles)

The route commences at the gate, which is the only entrance to the reserve. Follow the arrows slightly to the right, walking initially over sandy soil, which supports the long-stemmed form of the sea pink or thrift. Meadow saxifrage, yellow rattle and lady's smock all pepper the field with white, yellow or pink in their respective seasons. The path soon runs through a damp low-lying area, where there is a variety of sedges and rushes, meadowsweet, marsh thistle and many orchids.

After walking along the edge of the rushes, turn left at the post along another low-lying area, where ragged-robin and early marsh-orchids are found. Then cross the field, and so to the first foot-bridge. The bridge crosses the Ancaster Beck, which runs clear over a sandy bottom. Mallard frequent the banks, and the holes in the old willows provide homes for small mammals as well as birds. Brooklime grows at the water's edge. The next field has good cover on both sides, with dense scrub beneath the trees. This quiet area is often the best place to watch for birds, squirrels and foxes. Marsh thistles are frequent. Turn left at the post to enter the wettest and largest field, where the greatest numbers of orchids grow, sometimes in their thousands. Marsh-marigolds grow at the edge of the sedge area and provide cover for nesting snipe. In winter the far end of the field can become flooded and attracts coot and duck. Adder's-tongue, a scarce plant that is difficult to find, also occurs.

Turn left at the post near one of the several guelder roses found on the reserve to cross the field and you will again face the beck. Head towards the willows, walking through large patches of blunt-flowered rush and short herb-rich areas, where purple moor-grass, devil's-bit scabious and marsh valerian occur. Follow the water downstream past many alder trees, with water-cress, water forget-me-not and lesser water-parsnip at the water's edge. The second bridge is hidden among hawthorn bushes. Look for sticklebacks in the stream. On the other side, the gate at the end of the route will be seen. Walk towards it, looking for bugle in spring, and oxeye daisy and rough hawkbit in summer.

Moor Farm

Kirkby-on-Bain
OS: 122 GR: TF 226635 47.7 hectares (118 acres)
Freehold 1970
Map ref: 35

Location and Access

The reserve is situated to the north-east of Woodhall Spa, some 500 m north from the junction with the Woodhall Spa–Kirkby road. The entrance is on the west side of Wellsyke Lane. There is parking space inside the gate on the right. Visitors are requested to close the gate and keep to the waymarked route.

Description and Management

Moor Farm, which includes Rushbottom Close, is a mosaic of heath, dry and wet pastures, bog and woodland, on Fen-edge sands and gravels. Much of the woodland has developed from abandoned pastures during the last 80 years. The diversity of habitat supports a wide range of plants and animals, some of them now rare or localised in the Trust's area. Over 240 species of plant, 100 of bird, 24 of butterfly and 250 of moth have been recorded. The woodland is managed to develop maximum natural interest by selective thinning, while the bog area has been cleared of scrub, and measures are taken to retain a high water-level. The grassland areas are grazed or mown in a traditional manner in order to retain their existing characteristics. Adders occur in the reserve.

Rabbits on Moor Farm

Waymarked Route

Short route: 1.6 km (1 mile); long route: 2.5 km (1.5 miles)

From the parking area follow the waymarked route, branching right from the main track. Pass the gorse and woodland on the left where linnets may be seen in the summer and redpolls in winter. The field on the right is closely grazed and is noted for its Breckland-type flora, with cudweed, common centaury, common bird's-foot-trefoil, crane's-bill and stork's-bill. Pass between the pair of large European larch into the big field, which has many interesting plants including both silver hair-grass and early hair-grass, the uncommon creeping St John's-wort, shepherd's cress and basil thyme. Green woodpecker and mistle thrush are often seen here.

Cross the field along the path heading for the kissing-gate near the far corner. Through the gate the route turns left along the woodland edge, but a short diversion to the right leads to a hide giving views over Rushbottom Close, which is let for cattle grazing. The wetter parts have a tall herb community, including ragged-robin, purple-loosestrife, lesser spearwort, club-rushes and various sedges. Snipe and lapwing have nested in the area. There are often coot, moorhen and little grebe on the small pool, and several species of dragonfly.

Return along the woodland edge path. Opposite the kissing-gate there are views of wet heath and bog on the right, where there is an extensive area of purple moor-grass with patches of cross-leaved heath, common cottongrass and bog mosses. Tree pipits nest in this area, which is grazed by cattle at certain times of the year. Please do not go beyond the edge of the trees. Continue along the woodland edge path. At the junction with another track on the edge of the woodland there is a choice of routes. The short route to the left skirts the edge of the big field and rejoins the main route at the corner of the Heath Field (see below).

Turn right for the longer route, passing through the mixed woodland with birch, oak, alder and rowan. In autumn the wood is especially rich in fungi. Woodcocks are often to be seen in the wood throughout the year. The path leads to a brook, which marks the parish boundary, and turns left to follow it. The swift-flowing, clear water is full of aquatic plants and, in winter when the level is high, mallard, teal and grey wagtail may be seen. Continue until the path turns left into the birch woodland. This winding path crosses two wooden bridges over ditches marking old field boundaries and leads, via a stile and bridge, to the corner of the Heath Field.

The Heath Field has a wet heath flora with cross-leaved heath, devil's-bit scabious, heath spotted-orchid, lousewort and cottongrass. The waymarked route follows the edge of the field beside the fence. The field is grazed by cattle for part of the year. The route turns back into the woodland at the farther end of the Heath Field. This woodland is maturer than that seen earlier, with many large oaks and pines. All three species of woodpecker, spotted flycatcher, jay, treecreeper, and many other birds may be seen here. The path leaves the wood and winds through interesting woodland edge and scrub habitats with gorse and heather. This is a likely spot to see slow-worm, grass snake and adder. Soon the path comes to the main track only a short distance from your starting point. Turn right to the parking area.

Moor Farm

GR at E TF 226635

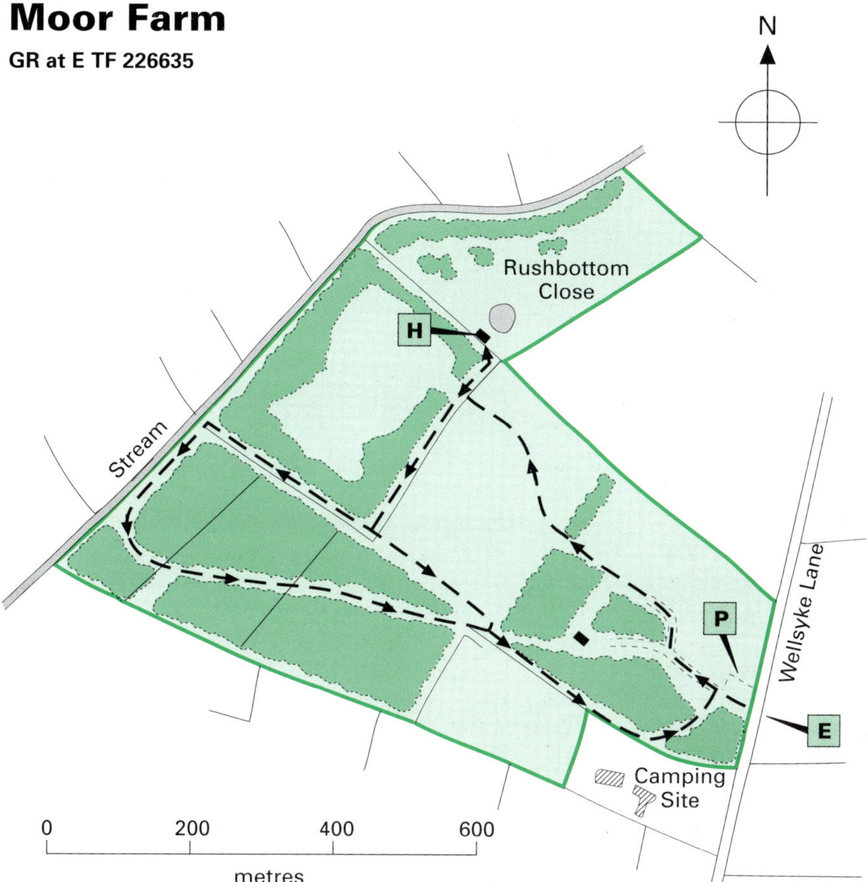

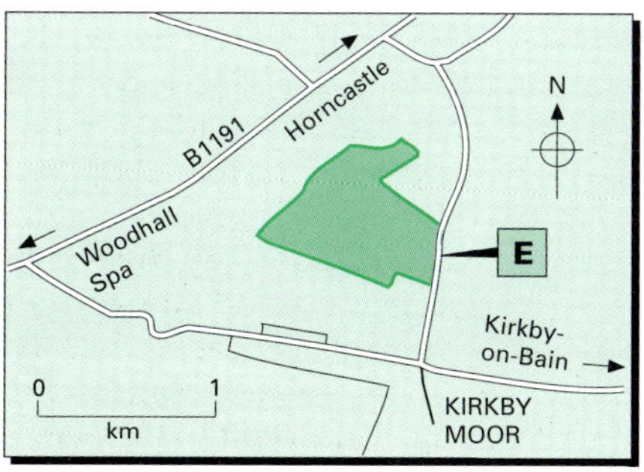

Moulton Marsh

Moulton
OS: 131 GR: SK 344336 to 326325 36.4 hectares (90 acres)
Leasehold 1979
Map ref: 99

Location and Access

The reserve is situated on the south bank of the River Welland below Fosdyke Bridge on the A17. Turn east off the A17 south of Fosdyke Bridge and at the end of this road turn towards the river bank. Follow the roadway to the reserve car park. From the car park, go past the gate with the reserve sign and follow the track to the observation hide overlooking the reserve. A path behind a screening bank leads to four other hides overlooking the scrapes; a fifth hide overlooks a shallow lagoon further on. Visitors are requested to keep to the paths.

Description and Management

The long narrow stretch of the reserve is entirely man-made, mostly as a result of soil extraction when the sea bank was raised. The area adjacent to the car park used to be a refuse tip and has now been covered with soil and planted with trees; this area covers 6 hectares (15 acres). Upstream (west) of this area are two large lagoons with islands, and adjacent to the river are shallow tidal scrapes, also with islands. Beyond this is an area of saltmarsh and creeks, followed by another lagoon. This whole area totals 14 hectares (35 acres). Finally there are some 16 hectares (40 acres) of saltmarsh.

While the number of bird species seen from the reserve stands at 120, only 25 species have been recorded breeding. In autumn and winter waders and wildfowl are present. Several short-eared owls are also usually to be seen in winter. In summer warblers, finches and buntings can be seen in the bushes. The sea bank alongside the reserve has a variety of wildflowers and attracts butterflies and insects. Strawberry clover and sea-milkwort, and two species of groundhopper, occur on the reserve.

Management consists of keeping the grass short in the scrapes to suit wintering wildfowl, and maintenance of the young trees on the newly planted refuse tip area. The water-level in the lagoons and scrapes is controlled by a series of pipes and flaps.

Wigeon

Moulton Marsh

GR at SK 344336–326325

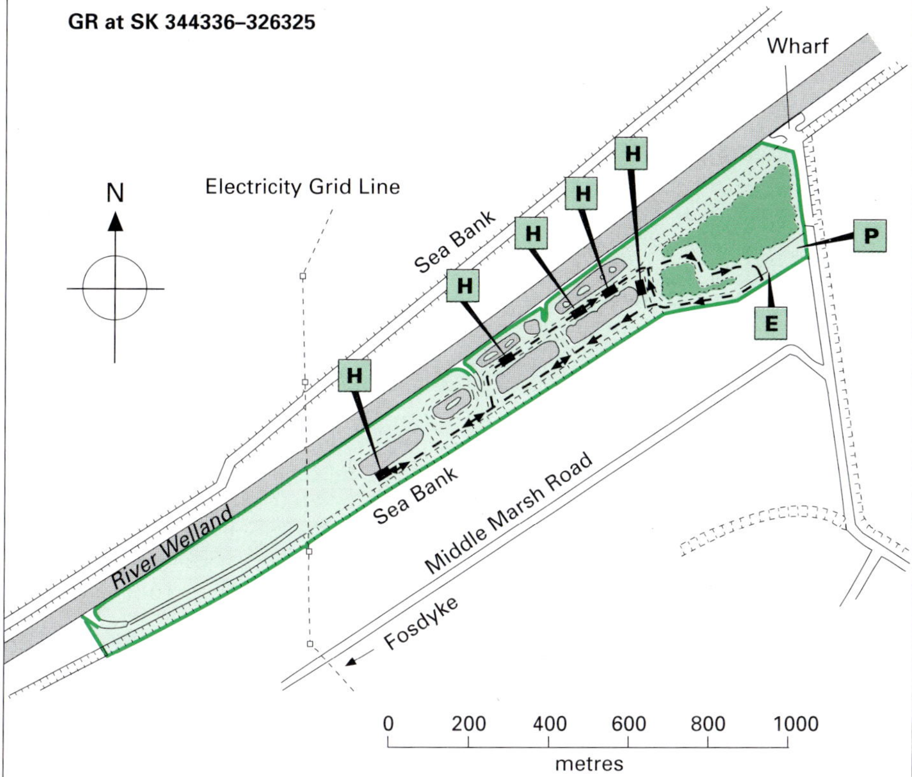

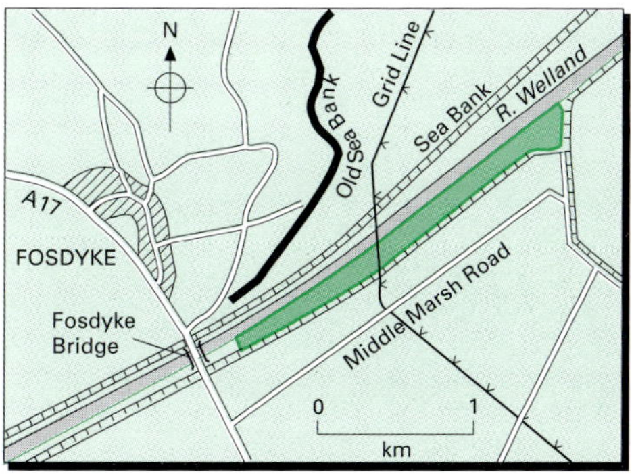

Muckton Wood

Muckton
OS: 122 GR: TF 382810 16.7 hectares (41.2 acres)
Freehold 1983
Map ref: 66

Location and Access

The wood is reached on foot by a 250-m access track, which leaves the Muckton–Belleau road about 1 km (0.6 miles) south of the village of Muckton at grid reference TF 381807. There is limited parking space on the road verge. Visitors are requested to keep to the rides and to go quietly.

Description and Management

This is an ancient semi-natural woodland on boulder clay soil on the eastern edge of the Lincolnshire Wolds. Oak and ash are the dominant trees with abundant field maple, aspen on the wood margins and alder in the wetter areas. The understorey is dense in places with some neglected hazel coppice, hawthorn, blackthorn and honeysuckle. Bramble tends to dominate the ground flora, but there are many woodland flowers, including wood anemone, woodruff, moschatel, bluebell, wood-sorrel and an abundance of herb-Paris. Early-purple orchid, broad-leaved helleborine and butterfly orchid are also to be seen, along with both the soft- and hard-shield ferns. The wood is attractive to many warblers, and breeding birds include great spotted woodpecker, treecreeper and tree sparrow.

When purchased in 1983, the wood was much neglected with a good deal of overstood hazel coppice and overgrown rides. Management so far has involved ride and ditch clearance, and bridge and pond construction. Traditional coppice-with-standards management has been re-established in parts of the wood.

Herb-Paris

Muckton Wood

GR at E TF 382810

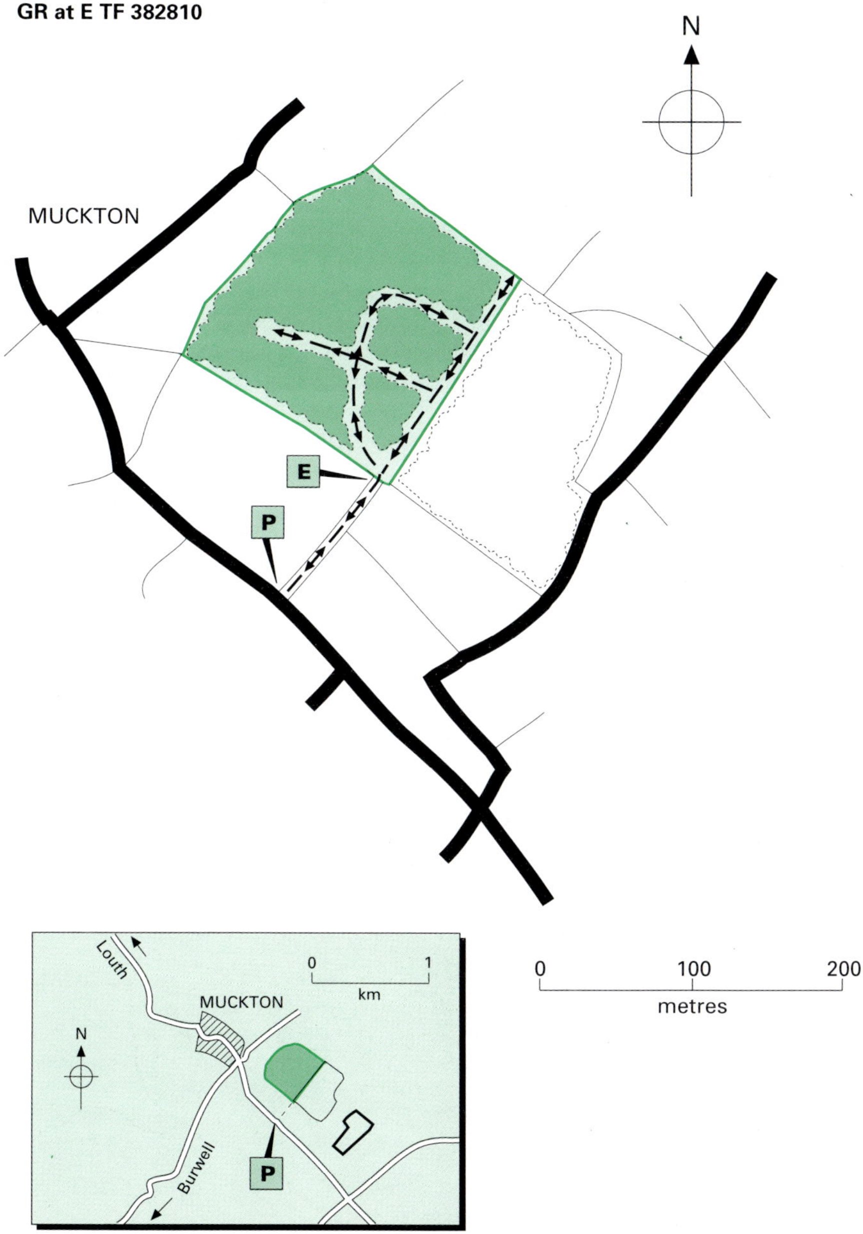

Pasture Wharf

Barton-upon-Humber
OS: 112 GR: TA 044229 (main entrance) 21.2 hectares (52.4 acres)
Freehold 1992
Map ref: 93

SSSI

Location and Access

The reserve lies on the Humber Bank between Barton and Barrow. Access is either from the Humber Bank footpath or from Pasture Road, Barton Waterside. Cars may be left at the small car park at grid reference TA 044229.

Description and Management

This important addition to the Trust's Humber Bank nature reserves consists of flooded clay pits, with open water and reedbeds, and 1.3 km (0.8 miles) of foreshore, including the old wharf. It is also the site of a former tile works (where a kiln chimney still stands). The mosaic of open water, reedbeds and intervening rough grassland and scrub provides excellent habitat for breeding birds, such as great crested and little grebes, bearded tit, reed warbler and blackcap. Mallard, pochard, tufted duck, gadwall and goldeneye occur in good numbers in winter. The foreshore, mudflats and flooded grassland attract many wading birds, including dunlin, redshank, grey plover and bar-tailed godwit. Bitterns occur in winter, and with careful management we hope they may return as a nesting species.

The aims of management are to retain a balance between open water and reed and to provide a sanctuary area by restricting access to some reedbeds and waterside areas. Reed-cutting and water-level control are also important.

Bearded tit

Pasture Wharf

GR at E TA 044229 (main entrance)

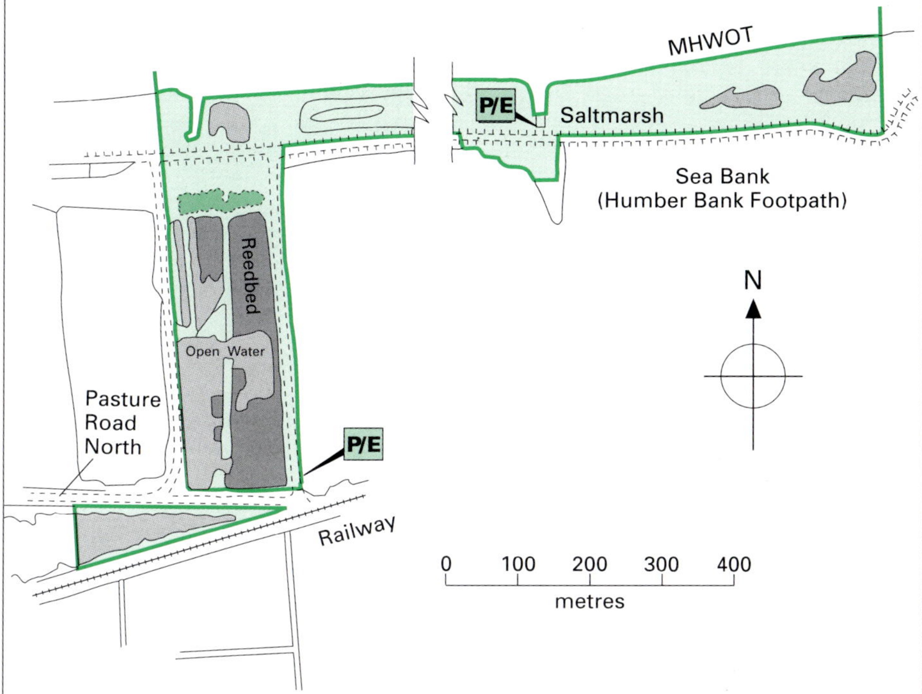

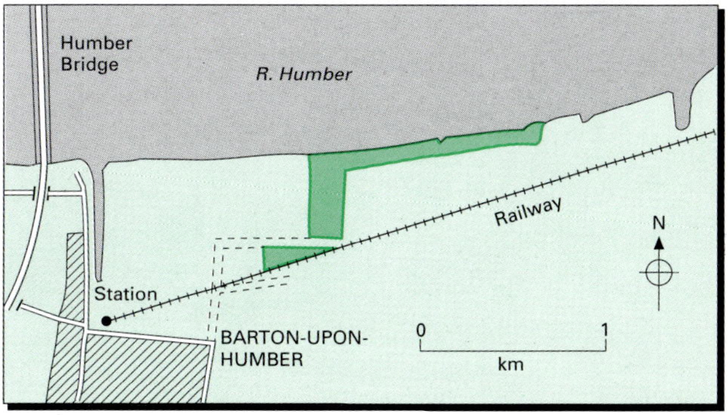

Pickering's Meadow

Dunholme
OS: 121 GR: TF 041793 2.8 hectares (6.9 acres)
Freehold 1984. Gift of Miss EH Pickering
Map ref: 70

Location and Access

The reserve is to the east of Dunholme on Ashing Lane, a green lane that runs from the village bypass. The meadow is on the left side after 1.5 km (0.9 miles). A public footpath crosses the reserve. There is suitable access for disabled persons via the main gate and along the footpath. Cars may be driven some way along the lane.

Description and Management

This attractive old meadow with tall hedges extends from Ashing Lane down to the Dunholme Beck, which is lined with tall crack willows. Meadow flowers include common bird's-foot trefoil, cowslip, tufted vetch, fleabane, hedge bedstraw and great burnet. There is also a large colony of adder's-tongue with cuckooflower. There is a field pond in the south-east corner with water figwort, reed sweet-grass, brooklime and a resident pair of moorhens. The hedges contain 13 different trees and shrubs, including hawthorn, blackthorn, field rose, dog rose, ash, dogwood, purging buckthorn and field maple. An owl box has been erected near the central dyke.

The meadow is traditionally hay cropped in mid-July and the aftermath is grazed. Management has involved clearing the pond, stock fencing, hedge laying and some tree planting with native deciduous trees and shrubs, such as oak and guelder-rose.

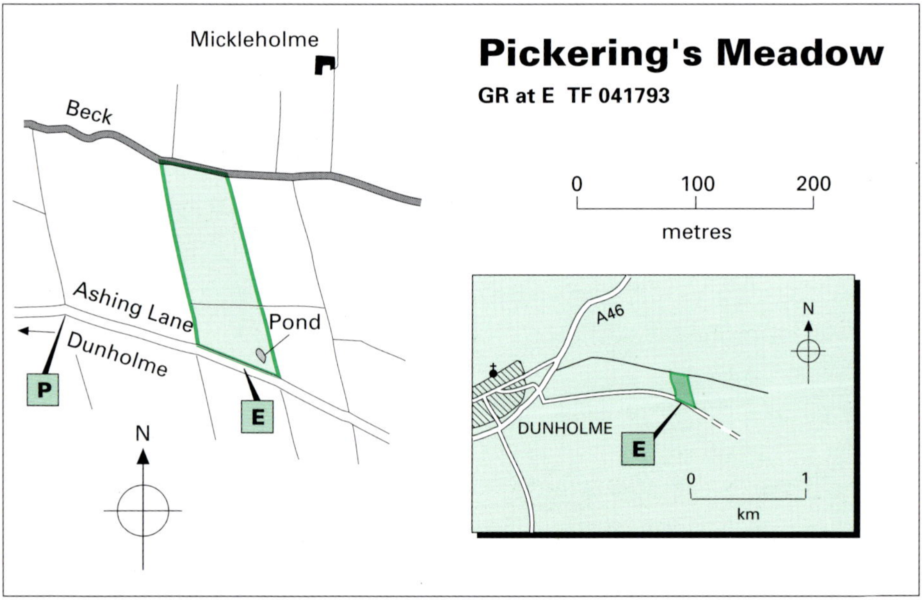

Rauceby Warren

South Rauceby
OS: 130 GR: TF 034441 8.8 hectares (21.9 acres)
Freehold 1968
Map ref: 36

SSSI

Location and Access

The reserve consists of a 1.3-km linear strip that runs along the south side of the A153 Sleaford–Grantham road, about 4 km (2.5 miles) west of Sleaford. There is parking space through a gate near the Sleaford end or on the open space at the western end.

Description and Management

The reserve lies on the sandy warrens overlying limestone formed by glacial melt waters, which carried sand through the Ancaster Gap. Few patches of the old heathland turf of these deposits now remain. Much of the reserve consists of abandoned sand and gravel pits, which were worked between 1920 and 1967.

The mixture of calcareous and acid soils gives rise to a varied flora rather like that of the Breckland of Norfolk and Suffolk. Over 320 plants have been recorded, including many localised species and one national rarity, the smooth rupturewort. Over 90 species of birds have been recorded of which 25 are known to nest. Records also include 60 species of aculeate wasps and bees.

Management consists largely of mowing grassland and tall herb vegetation, preventing scrub invasion, planting trees and repairing the stone wall on the northern boundary.

Smooth rupturewort

Waymarked Route

Short route: 1.3 km (0.9 miles); long route: 2.4 km (1.5 miles)

Before starting on the waymarked route, a close examination of the 50 or so species of plants in the car park is recommended. These include hoary cinquefoil, common cudweed and the rare smooth rupturewort. Then turn right towards the western end of the reserve and walk along the northern boundary. In summer there is a fine display of viper's-bugloss. In winter look out for stoats hunting the rabbits along this stretch. The short route turns left to the railway, but the long route continues along the wall side. Keep a look out for grass snakes and common lizards. Twenty-two species of butterflies have been recorded. Common blues are numerous in May and August.

Follow the bed of the old quarry railway. The bank is covered with flowering plants, including dwarf spurge, small toadflax, Canadian and blue fleabane and mignonette. Look also for bird's-foot trefoil, knotted pearlwort, scarlet pimpernel and smooth rupturewort.

At the end of the lagoon, cross over to the railway side and turn eastwards. There are displays of purple milk-vetch, tuberous pea, field scabious and toadflax to be seen. Continue along the fence, rejoining the short route on the way.

In the tangled vegetation at about the centre of the reserve, mallard, red-legged partridge, reed bunting and pheasant nest. Spotted flycatcher, various species of finches and tits, whitethroat, willow warbler and many other birds are to be seen along the route. Woodcock sometimes winter in the area.

When reaching the end of the hawthorn scrub, turn left across the springy, rabbit-grazed, mossy turf to the edge of the eastern lagoon, which retains water longer than any other. Tufted duck, pochard and little grebe are generally present in the winter and early spring, and Canada geese successfully nest on the banks of the lagoon. Great crested newts also breed here. To avoid undue disturbance to the birds, and if the water-level is low enough, turn left through the willows and cross the centre of the lagoon to the car park. On this part of the route, spectacular displays of evening primrose are to be seen from July onwards, though the reserve management plan includes the progressive removal of this alien species.

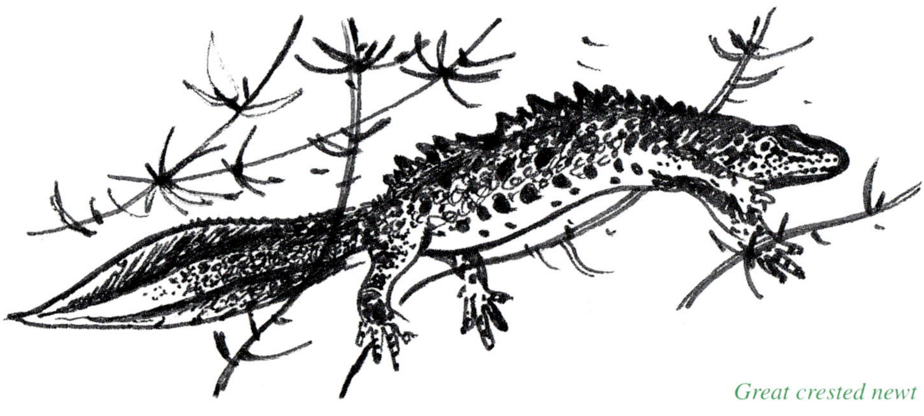

Great crested newt

Rauceby Warren

GR at E TF 034441

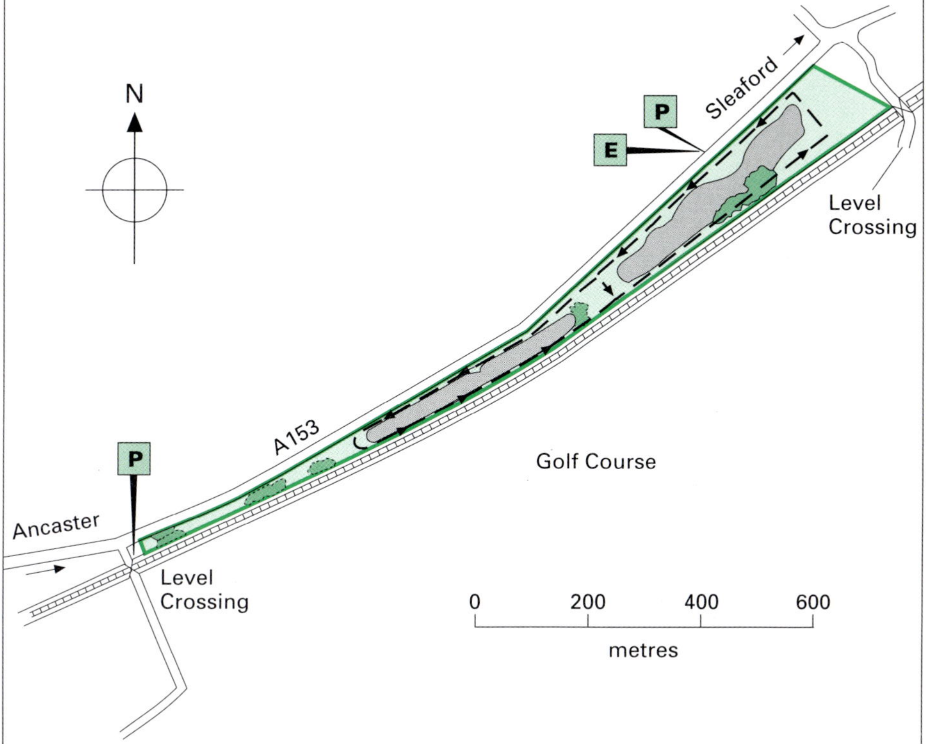

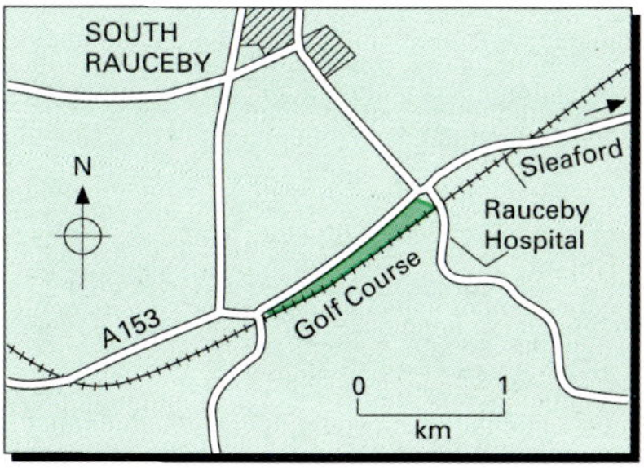

Red Hill

Goulceby
OS: 122 GR: TF 264806 4.3 hectares (10.6 acres)
Freehold 1967
Map ref: 37

SSSI

Location and Access

The reserve adjoins the Goulceby–Raithby road, about 1.5 km (0.9 miles) south-west of the Bluestone Heath Road. Parking places are provided at the top end and to the east of the reserve. A leaflet about the reserve is available from the Trust's headquarters in Horncastle.

Description and Management

The area comprises 1.6 hectares (4 acres) of steep chalk escarpment grassland with some scrub and 1.6 hectares (4 acres) of old plateau grassland, and a disused quarry with a famous exposure of Red Chalk, which is rich in fossils, particularly belemnites and brachiopods. Below the Red Chalk there is a considerable thickness of Carstone, here a coarse, pebbly sandstone, and above it a thin capping of the white Lower Chalk.

The plateau grassland is probably an ancient fragment of Lincolnshire Wold downland, and there is a rich assemblage of chalk plants here and in the quarry, including localised species such as felwort, yellow-wort, basil thyme, kidney vetch and pyramidal and bee orchids. Owing to the dominance of tor-grass the hillside vegetation is less varied.

The commoner grassland butterflies and moths, including an abundance of six-spotted burnet moths, are present. There are scarce insects among other orders and an exceptionally high spider population. The meadow pipit, a localised breeding species inland in Lincolnshire, nests in the reserve. Common lizard and grass snake occur.

Some mowing of grassland and verges is undertaken, but when possible the hillside and plateau grassland are grazed by sheep to maintain floristic diversity. The chalk cliffs are friable so that erosion easily occurs: they should not be climbed. For the same reason some areas are fenced off in order to reduce wear.

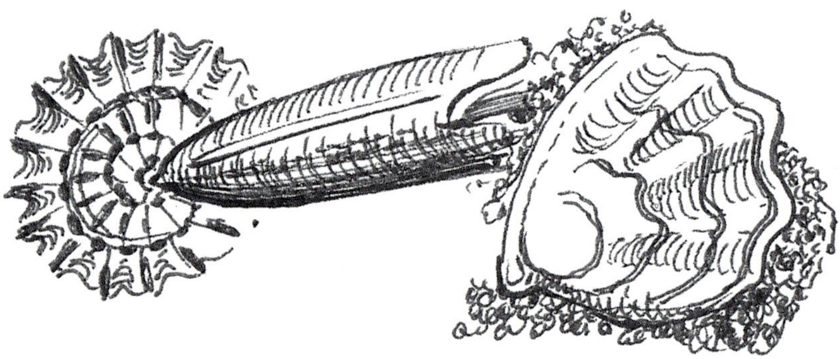

A collection of fossils

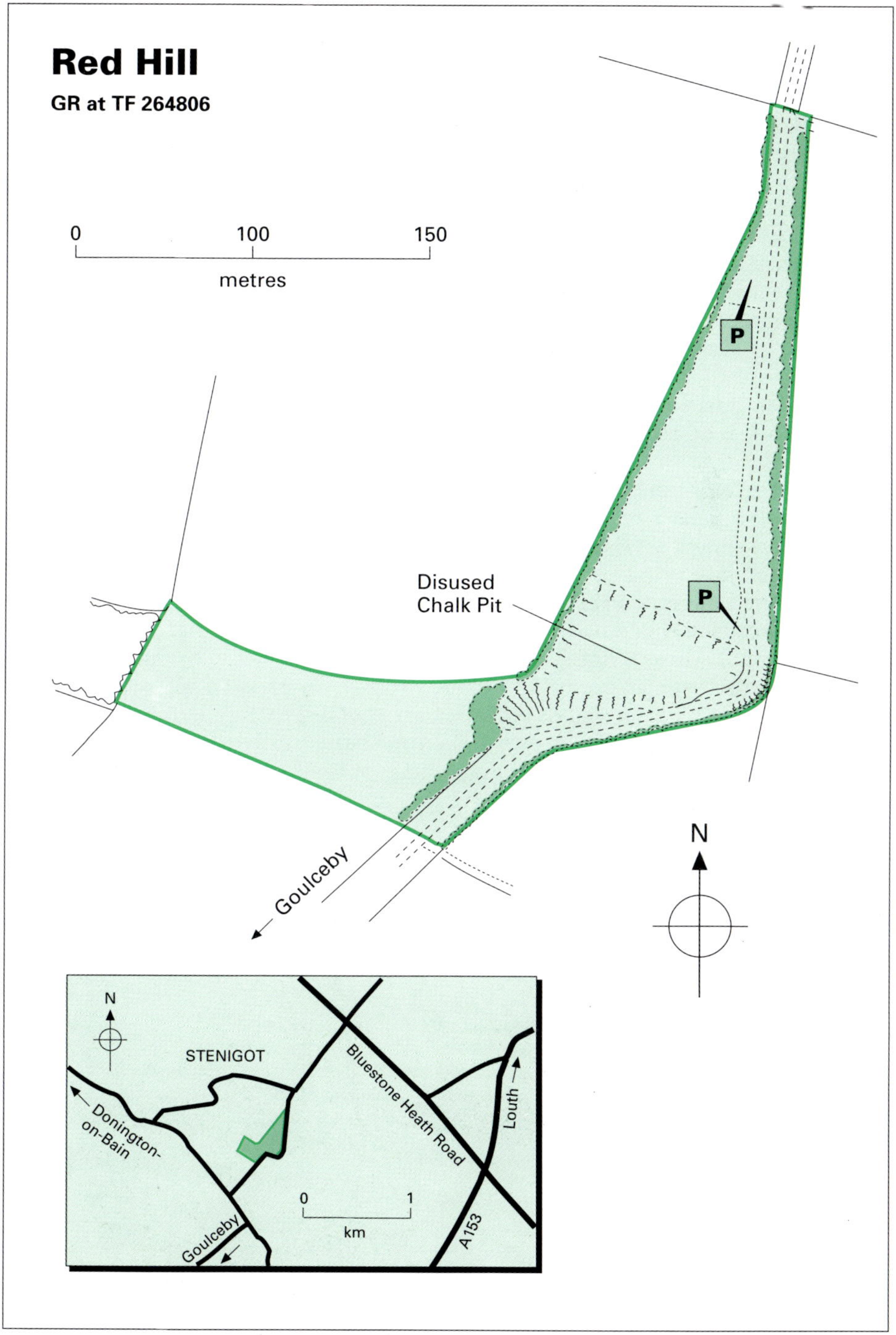

Red Hill
GR at TF 264806
0
100
150
metres
Disused
Chalk Pit
P
P
Goulceby
N
N
STENIGOT
Bluestone Heath Road
Louth
Donington-
on-Bain
0
1
km
A153
Goulceby

Rigsby Wood

Rigsby-with-Ailby
OS: 122 GR: TF 421762 15.1 hectares (37.3 acres)
Freehold 1980
Map ref: 38

Location and Access

The wood lies at the foot of the Wolds, about 3 km (1.8 miles) west of Alford. Turn north towards South Thoresby at the Miles Cross Hill crossroads on the A1104 road half way between Ulceby Cross and Alford. After 2 km, the wood will be seen set back from the road on the right. Cars may be parked on the roadside. The wood is reached on foot along a short field track.

Description and Management

Rigsby Wood is an ancient woodland lying partly on chalky Boulder Clay and partly on glacial sands at the edge of the Wolds. Although some areas have been replanted, the reserve retains a rich ground flora, and a substantial part is old coppice of hazel and ash with standard trees of oak and ash. The path passes through a narrow section of the wood known as The Pingle in which old ridge and furrow can be detected, showing that this is not part of the ancient wood. The woodland soon opens out and the path crosses the medieval boundary bank into the old part of the wood. There is old plantation woodland on the left and old coppice on the right. In this older part of the wood, bluebells in abundance, wood anemone, wood-sorrel, early-purple orchid, sweet woodruff and ragged-robin, which is an old meadow relict species, are to be seen. Along the old boundary between Rigsby and Haugh parishes, there is a fine medieval bank, and another one between Rigsby and Ailby Woods with large field maples. There is a good range of breeding birds, including tawny owl, great spotted woodpecker, treecreeper, finches and five species of both tits and summer warblers. Barn owls can occasionally be seen hunting along the rides. In winter woodcock, goldcrest and redpoll are frequent visitors.

Management is aimed at perpetuating primary woodland with both coppice and high forest conditions. The main ride has been considerably widened.

Wood anemone

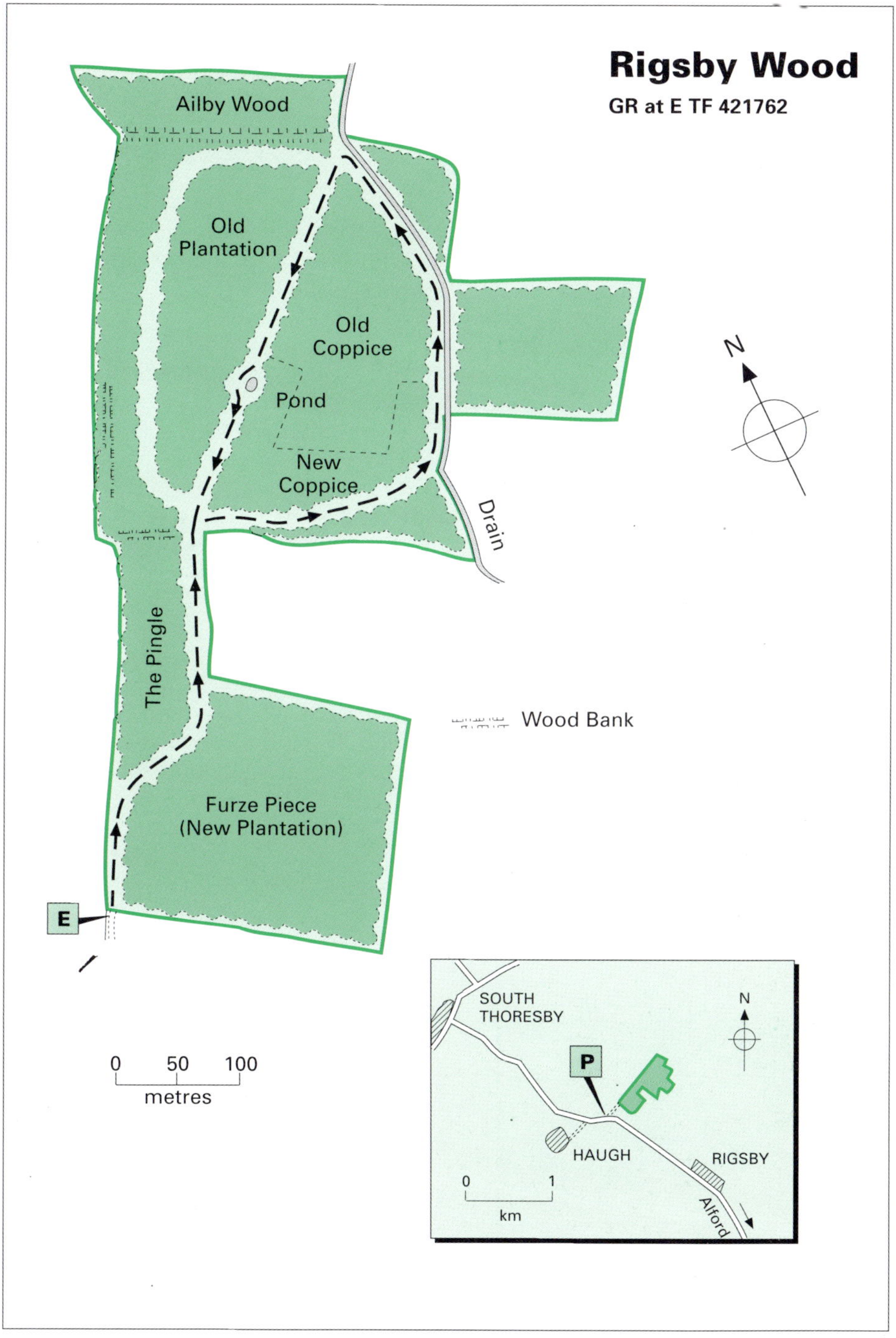
Rigsby Wood
GR at E TF 421762
Ailby Wood
Old Plantation
Old Coppice
Pond
New Coppice
Drain
The Pingle
Furze Piece (New Plantation)
E
Wood Bank
N
0 50 100
metres
SOUTH THORESBY
P
HAUGH
RIGSBY
Alford
0 1
km

Robert's Field

Careby, Aunby and Holywell
OS: 130 GR: TF 000150 4 hectares (10 acres)
Leasehold 1991
Map ref: 75

Location and Access

The reserve consists of a small meadow and restored limestone grassland on the east side of the minor road from Holywell to Pickworth close to the county boundary. Access is via a broad track at a wooden gate.

Description and Management

The Trust has had a 1-acre meadow reserve here, under a management agreement with the Forestry Commission, since 1985. The remainder of the site, which was formerly limestone grassland, was planted with conifers in the 1960s despite its SSSI status and the Trust's efforts to save it. The remaining portion of a long lease was purchased from the Forestry Commission in 1991.

Robert's Field was once the finest butterfly site in Lincolnshire, with at least four species now extinct in the county. The surviving grassland retains many characteristic limestone plants, including pyramidal orchid, clustered bellflower, autumn gentian and rock-rose. There are still a few of the grassland butterflies and it is a noted inland site for the green hairstreak.

The Trust embarked on a plan to restore Roberts's Field to limestone grassland in 1994. This is an ambitious plan that will take many years to complete. Already the conifer trees have been felled, the roots have been removed, and the ground has been levelled and re-seeded with hay seeds from local limestone grassland. Eventually the whole site will be stock fenced so that the grassland can be managed by autumn grazing with sheep.

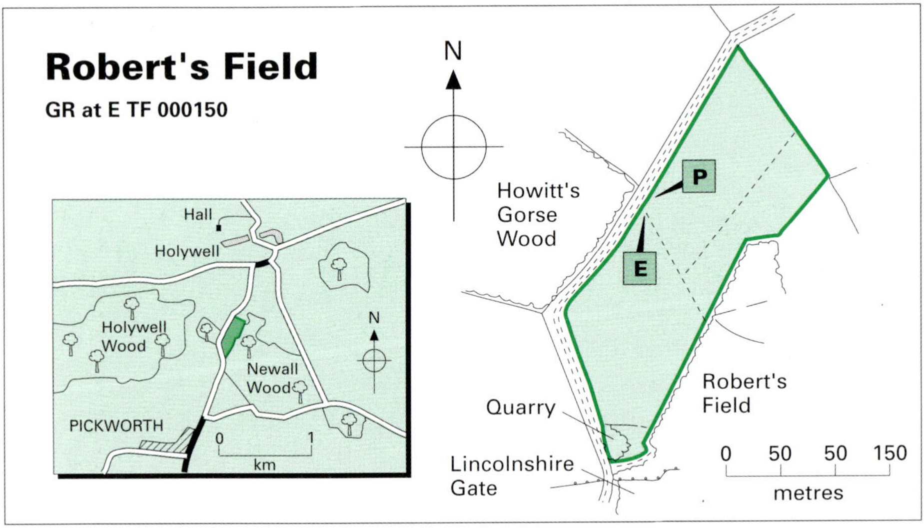

Rosper Road Pools

South Killingholme
OS: 113 GR: TA 172168 9.8 hectares (24.3 acres)
Leasehold 1980
Map ref: 39

Location and Access

The reserve is situated about 2 km (1.2 miles) north-west of Immingham, just inland from South Killingholme Haven. Take the A160 road from South Killingholme towards the Haven. Shortly after a roundabout is a railway bridge followed by a turning to the left. This is Rosper Road and the reserve is a short distance on the right. No access is possible to the reserve but it may be easily viewed from the road.

Description and Management

Consisting of a flood-relief reservoir with associated wet grassland, the reserve has become of great ornithological interest in recent years. Ninety bird species have been recorded. Teal, which are abundant in winter, often stay late into spring. Pintail and garganey are frequent visitors. The site is particularly good for wading birds, the greatest variety of which are to be seen in autumn when muddy margins suitable for feeding are exposed. Snipe, little stint, wood and green sandpipers, spotted redshank and greenshank are all regular visitors. Up to 70 ruff have been recorded in winter, and in some years a few remain to display in spring, along with migrant black-tailed godwits. Lapwings sometimes nest in the marshy grassland.

Grazing to provide optimum conditions for migrant and nesting birds is the main management aim.

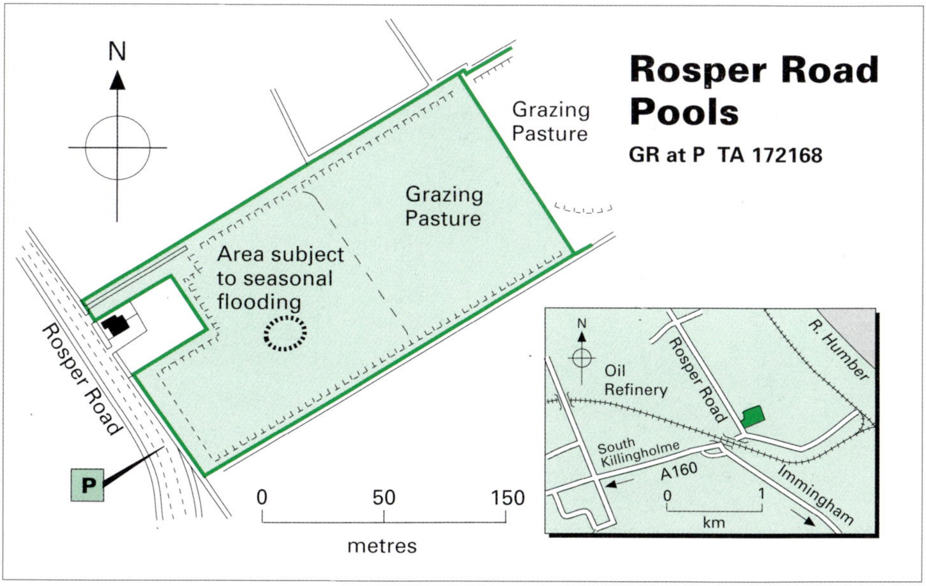

Roughton Moor Wood

Roughton
OS: 122 GR: TF 212632 10.1 hectares (25 acres)
Freehold 1988. Part gift from Mr GJ Tugwell
Map ref: 90

Location and Access

The reserve lies to the north of Moor Lane, which runs from Woodhall Spa to Kirkby-on-Bain. The only access is from Moor Lane. There is a roadside parking area a short distance towards Woodhall Spa. Cars should not be parked where they will obstruct the entrance to the poultry unit. The reserve is entered by a stile. Visitors are requested to keep to the waymarked route.

Description and Management

The reserve consists of woodland of oak, birch, rowan and Scots pine. The woodland has regenerated on former heathland, some remnants of which still survive in open glades. There are some fine trees and there is an impressive avenue of ageing Corsican pines planted apparently as the approach to a large house, which was never built. The wood has a varied bird population, including great spotted woodpecker, treecreeper, coal tit and goldcrest. Fungi are also a notable feature. The reserve is a valuable adjunct to the nearby Kirkby Moor and Moor Farm reserves.

Management is designed to maintain the woodland by natural regeneration as far as possible. Where heather and other open heath plants survive, glades will be kept open and some will be widened.

Fly agaric

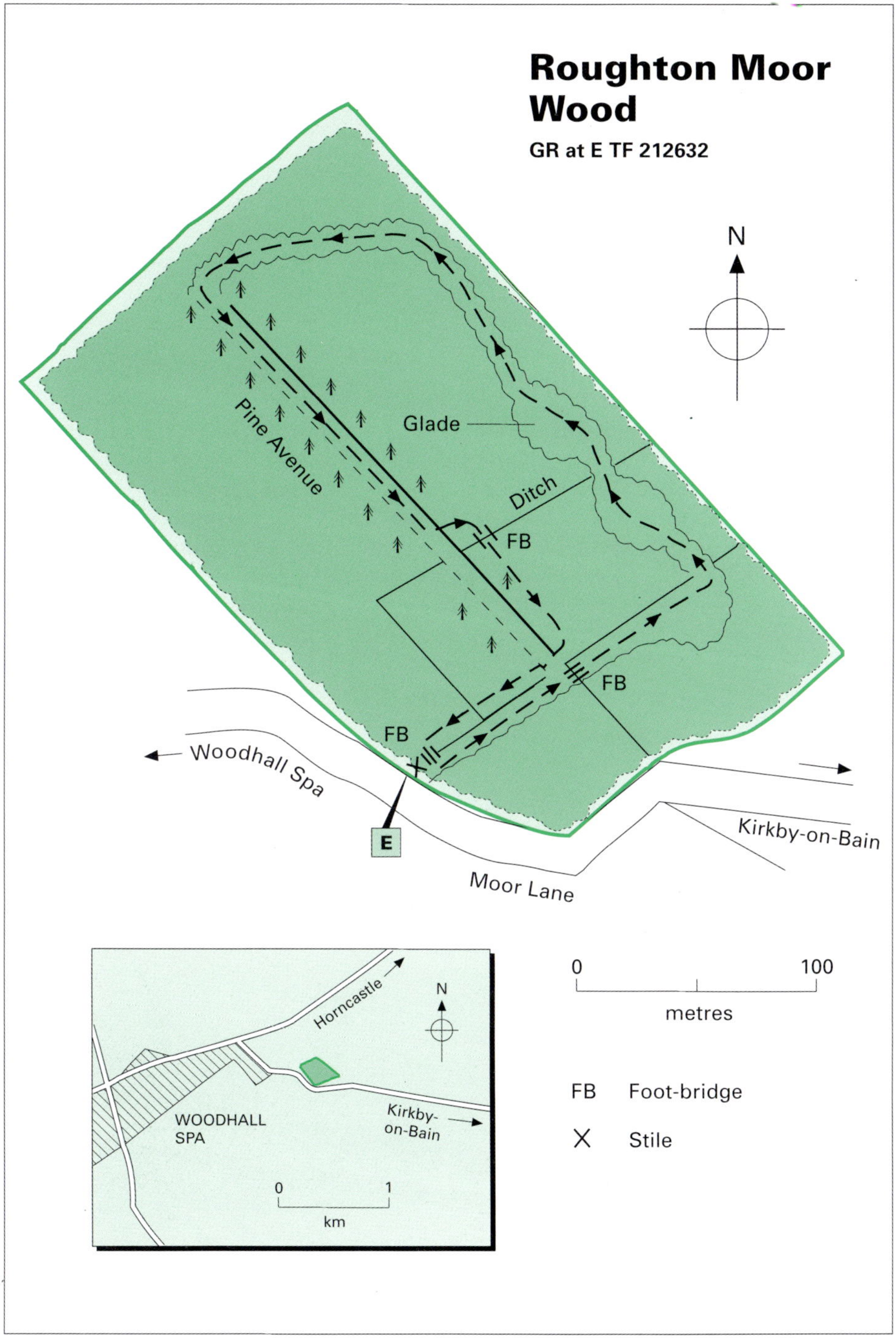
Roughton Moor Wood
GR at E TF 212632
N
Glade
Pine Avenue
Ditch
FB
FB
FB
E
Woodhall Spa
Kirkby-on-Bain
Moor Lane
Horncastle
N
WOODHALL SPA
Kirkby-on-Bain
0
1
km
0
100
metres
FB Foot-bridge
X Stile

Rush Furlong

Haxey
OS: 112 GR: SE 781004 0.4 hectares (1 acre)
Freehold 1978
Map ref: 40

Location and Access

The reserve lies to the east of the Isle of Axholme's main road, the A161, just north of Haxey. Follow the tracks as shown on the map.

Description and Management

Rush Furlong is a fragment of the once extensive system of strip farming in the Isle of Axholme that has by chance survived. The nearby land is now mostly arable, but this small meadow grassland strip has been managed only by hay-cutting for as long as can be traced. This regular, unchanging management and lack of disturbance, combined with a soil derived from the base-rich Keuper Marl, has resulted in a rich flora. Over 60 species of flowering plants have been recorded, including yellow rattle, oxeye daisy, rough hawk's-beard, agrimony, meadow oat-grass and primrose. The speciality of the site is the uncommon green-winged orchid of which several hundred spikes are to be found in most years. Adder's-tongue also occurs.

The management aim is to continue the traditional practice of hay-cutting.

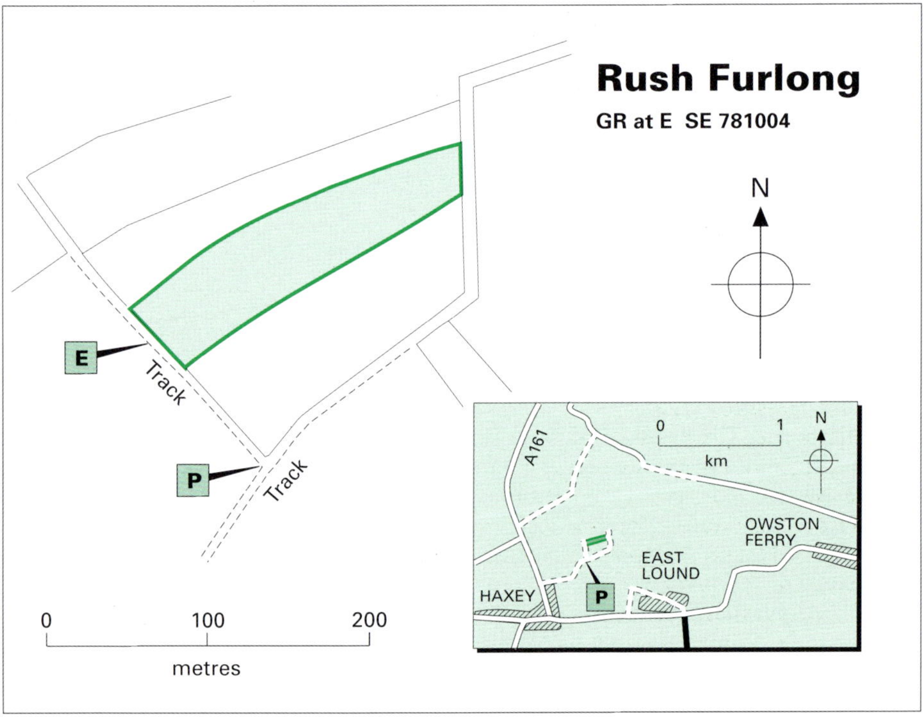

Saltfleetby–Theddlethorpe Dunes National Nature Reserve

Saltfleetby and Theddlethorpe
OS: 113/122 GR: TF 467917 (Rimac entrance)
Trust: 43 hectares (106.5 acres) NNR: 440 hectares (1090 acres)
Part freehold, part leasehold, part management agreement 1968
Map ref: 41

SSSI

The larger part, 440 hectares, is a statutory National Nature Reserve managed by English Nature. Approximately 38 hectares (95 acres) are managed by the Trust on a licence agreement with the Ministry of Defence. Another 4 hectares (9.8 acres) are leased from Lincolnshire County Council, and the Trust owns about 3 hectares (7.5 acres) comprising the Sea View Field and Rimac Bridge. The whole area is managed in close cooperation between the Trust and English Nature.

Location and Access

The Saltfleetby–Theddlethorpe reserve occupies over 7 km (4.4 miles) of coast between Mablethorpe North End in the south and Saltfleet Haven in the north. There are six main access points from the A1031 coast road at (from north to south): Paradise (GR: TF 456934), Sea View (GR: TF 465924), Rimac (GR: TF 467917), Coastguard Cottages (GR: TF 478901), Brickyard (GR: TF 483893) and Crook Bank (GR: TF 489882). There are parking places at each of these entrances. Although most of the reserve is open to the public, visitors are requested to keep out of the signed sanctuary areas.

An easy access trail, with display boards, has been constructed at Rimac to allow people of less mobility to see much of the habitat described. A free interpretive leaflet is available in the Rimac car park or from the Trust's headquarters in Horncastle. For those who are visually impaired, a free audio tape is available from the Trust's headquarters.

Description and Management

This important reserve contains tidal sand and mudflats, salt and freshwater marshes and sand dunes. On the foreshore, accreting mud and silt flats and saltmarsh in the north give way to a narrower sandy beach at the southern end. The sand dunes are also much wider in the north and there is an extensive freshwater marsh between two dune ridges, which converge into a narrower ridge south of Churchill Lane at Theddlethorpe. The much older landward dunes developed on a storm beach formed in the 13th century. The second ridge on the northern half of the reserve, enclosing the freshwater marsh, developed in the mid-1800s following the diversion of the Great Eau. New dunes are now forming along the southern half of the reserve.

In July and August, the saltmarsh is striking as it is covered with an abundance of sea-lavender. Other plants include thrift, sea aster and the silver-grey-leaved sea-purslane. Extensive samphire beds occur at the seaward accreting side of the marsh. In places at the top of the marsh freshwater species are beginning to colonise. Visitors are advised to be aware of the large creeks in the northern saltmarsh and the danger of rising tides flooding up the creeks; for your safety it is important to check the local tide table.

The old sand dunes support a rich flora, especially where grazing keeps a short turf. A wide range of grasses occur with sand and glaucous sedge, and there are many colourful flowers, including autumnal hawkbit, bird's-foot trefoil, common centaury, field mouse-ear, viper's-bugloss and calcicolous plants such as felwort, fairy flax, carline thistle, and pyramidal and bee orchids. On a few slopes there are patches of the rare sand-dune form of lesser meadow-rue. In the absence of grazing, false oat-grass followed by scrub of sea-buckthorn, elder and hawthorn become dominant. The foredunes are colonised by sand couch and stabilised by marram, sea lyme-grass and red fescue.

The freshwater marsh is of an unusual kind, termed 'maritime fen'. Greater pond-sedge, sea club-rush and common reed are the dominant plants. However, in more open areas and along edges where the vegetation is shorter, there is a great variety of plants, including kingcup, lesser spearwort and yellow flag among the more colourful, with rare or localised species, such as marsh arrowgrass, lesser water-plantain, bog pimpernel, skullcap, great water-parsnip and marsh pea. Marsh helleborine also occurs. In many years a spectacular display of early and southern marsh-orchids and their hybrids can be seen in June.

The commoner grassland butterflies are all found on the reserve, and green hairstreaks are a notable feature of early summer. Moths include a range of marsh and saltmarsh species, some of which are scarce. Dykes and ponds support a variety of aquatic creatures, including dragonflies, beetles, boatmen, skaters and the water spider. Frogs, toads and smooth newts all abound, but the most notable of the amphibians is the rare nocturnal natterjack toad for which this is one of only two Lincolnshire localities. A variety of birds breed, including shelduck, redshank and nine types of warbler. The greatest variety of birds occurs at migration, while in winter birds of prey are attracted by the large numbers of waders feeding on the open shore and berry- and seed-eaters flocking to the shrublands and saltmarsh.

Management has concentrated on re-establishing traditional grazing and hay-cutting regimes to much of the dunes and freshwater marsh. Most of the old dyke and pond network has also been re-habilitated and some new ponds have been dug. This work has been essential in order to maintain the abundant wildlife of the open grasslands and the freshwater marsh. Several acres of sea-buckthorn have been removed from the old dune ridge to allow the rich grass sward to recolonise.

Natterjack toad

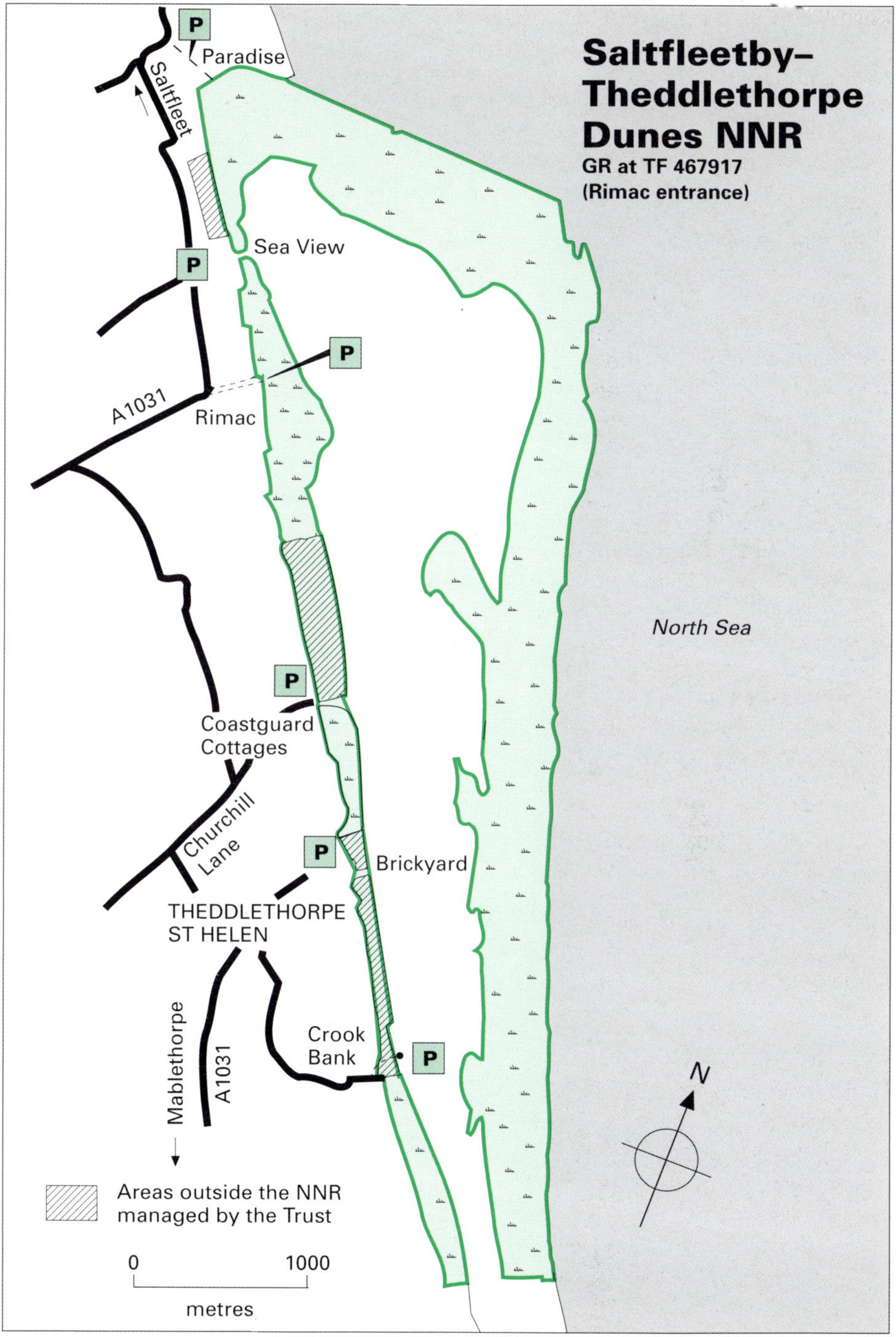
Saltfleetby–
Theddlethorpe
Dunes NNR
GR at TF 467917
(Rimac entrance)
P
Paradise
Saltfleet
Sea View
P
P
A1031
Rimac
North Sea
P
Coastguard
Cottages
Churchill
Lane
P
Brickyard
THEDDLETHORPE
ST HELEN
Crook
Bank
P
Mablethorpe
A1031
N
Areas outside the NNR
managed by the Trust
0
1000
metres

Sandilands Pit

Mablethorpe and Sutton
OS: 122 GR: TF 531802 1.5 hectares (3.8 acres)
Leasehold 1959
Map ref: 43

SSSI

Location and Access

The most northerly of the Sea Bank Clay Pits (see also Huttoft Bank Pit, Wolla Bank Pit and Chapel Pit entries), the reserve lies on the west side of the coast road just south of Sandilands. From Sutton-on-Sea follow the signpost to Sandilands. Pass Sandilands Golf Club on your left and the reserve will be found on the right. Cars may be parked in the access track (next to Stymie Cottage), but care should be taken not to obstruct the field access. For a location plan with adjoining reserves see page 105.

Description and Management

Excavated for clay to repair the sea bank after the 1953 floods, the flooded pit now supports many duck, especially in winter when snipe are also common. The fringing reeds and scrub provide good nesting habitat for reed and sedge warblers. There are beds of reed sweet-grass, sea couch and sea club-rush. Parts of the grassland are mown each autumn.

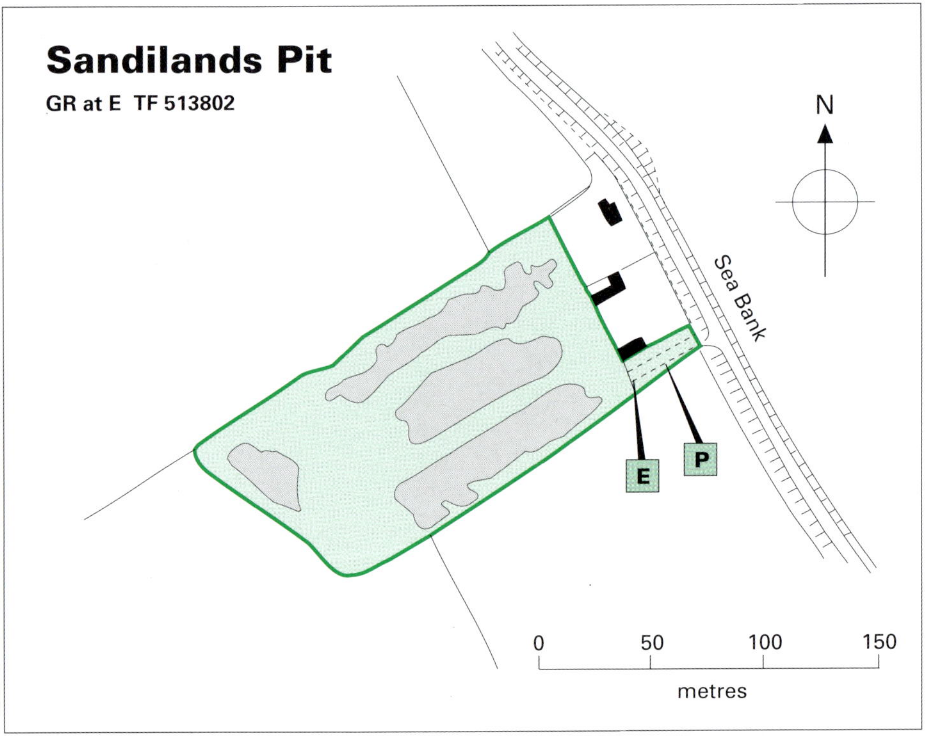

Scotton Common

Scotton
OS: 112 GR: SK 873985 (main entrance) 31.7 hectares (78.8 acres)
The Joe Duddington Memorial Reserve. Freehold 1954
Map ref: 42

Location and Access

The reserve lies on the left-hand side of the road leading to Scotton village from the A159 Gainsborough–Scunthorpe road. There are two entrances to the reserve. Some 300 m from the A159, on the left and just past the nursery, is a small footpath with restricted parking. Please avoid parking on the protected roadside verge. A further kilometre (0.6 miles) down the road is the main entrance where a gate gives access to a car park. Once on the reserve, please take special care during the nesting season. Adders occur in the reserve. There is a high fire risk during the summer months.

Following Joe Duddington's death in 1992 at the age of 81, the reserve was dedicated to him. Joe was a well known and loved local lepidopterist. There is a memorial stone erected in his memory near the car park.

Description and Management

This low-lying tract, the largest remaining fragment of the once extensive heathlands of Scotton Common, is covered in its drier parts by ling and by developing oak, birch and pine woodland. At the eastern end of the heath, on ground that was ploughed in the Second World War, wavy hair-grass is dominant. In wetter parts there is much purple moor-grass and cross-leaved heath; rarer species, such as bog asphodel and marsh gentian, also occur. To save it from destruction on a nearby site, the shrubby bog myrtle has been established in the reserve. Although some species of butterfly have become scarce or even disappeared in recent years, the list of butterflies and moths is impressive and includes many scarce and localised species, including a thriving population of emperor moth. Sandy banks attract solitary mining bees. Breeding birds include tree pipit and woodlark. Adders and common lizards are numerous.

The principal management problems, as on most of our heathland reserves, are to maintain open heath by preventing

Marsh gentian

encroachment of birch and pine scrub, and to halt the loss of wet heath species, such as sundews, which are suffering from a falling water-table. Since 1993 some 4 hectares (10 acres) of birch and pine have been felled. The whole reserve is now enclosed with sheep fencing, and grazing with Hebridean sheep is being carried out at appropriate times to prevent birch re-growth. It is hoped that tree removal will make the reserve wetter again. Nightjars have not bred for some years, but it is hoped that they may return.

After the recent felling there is no defined waymarked route over most of the reserve at present, although access to all parts of the reserve is easy. A new route will be made in due course. Until that time particular care is needed to avoid nesting birds.

The long verges adjoining the reserve are among the protected roadside verges scheduled by the County Council and the Trust (PRV 015). While visiting the reserve, walk along the verge bordering the reserve where you will see vetches, clovers, marsh thistle, comfrey and twayblade among the many grasses. On the opposite side of the road look for bog pimpernel in the ditches, and ferns and mosses on the banks. Many butterflies are to be seen, including brimstone, ringlet and skippers.

In 1988 an area of heathy fields of around 16 hectares (40 acres) to the north of the old reserve, known as Scotton Beck Fields, was purchased by the Trust. This area, which is reached by turning right off the access track between the car park and the road via a bridge and stile, is rich in heathland plants, including ling, cross-leaved heath and marsh gentian. The beck and streamside support bur-reed, floating sweet-grass, fool's water-cress, water mint and water forget-me-not. The old meanders contain kingcup, cuckooflower and marsh violet.

Hebridean sheep

Scotton Common

GR at E SK 873985 (main entrance)

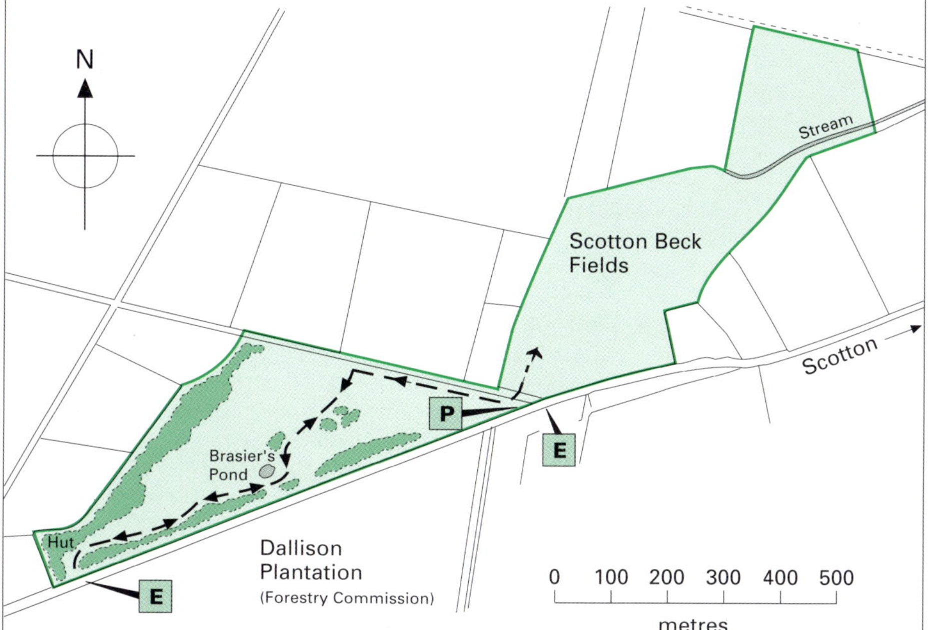

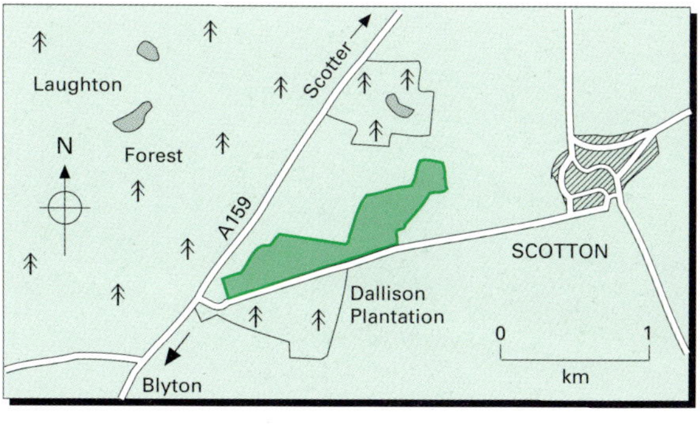

Sedge Hole Close

East Lound
OS: 112 GR: SK 793996 0.9 hectares (2.3 acres)
Freehold 1988/90
Map ref: 26

Location and Access

Access is by an unmetalled track that runs eastwards from East Lound village. The reserve is on the right after about 250 m. Cars may be parked on the roadside in the village. In the previous edition of the handbook this reserve was referred to by the name 'Seggholme Meadow'.

Description and Management

A small damp meadow originally divided into three strips by two hedgelines, of which remnants survive. There are approximately 110 species of plants in total. Typical old meadow plants include cowslip, great burnet, lesser knapweed, oxeye daisy and cuckooflower. There are also five species of buttercup, adder's-tongue, Good Friday-grass, ragged-robin and twayblade. In the south-west corner old retting pits, where flax used to be soaked, have become colonised by willows, reed sweet-grass and bittersweet. There are fine old boundary hedges with elm, hawthorn and field maple. The meadow area of the reserve, which accounts for half of the site, is managed as a traditional hay meadow and is enclosed for grazing purposes. The retting pits and reedbed provide a nesting place for birds. There is no path through this area but it can be readily observed from the meadow or adjacent footpaths (see map). The site is rich in butterflies and moths. The chimney sweeper moth, for instance, is abundant in the area when the pignut is in flower.

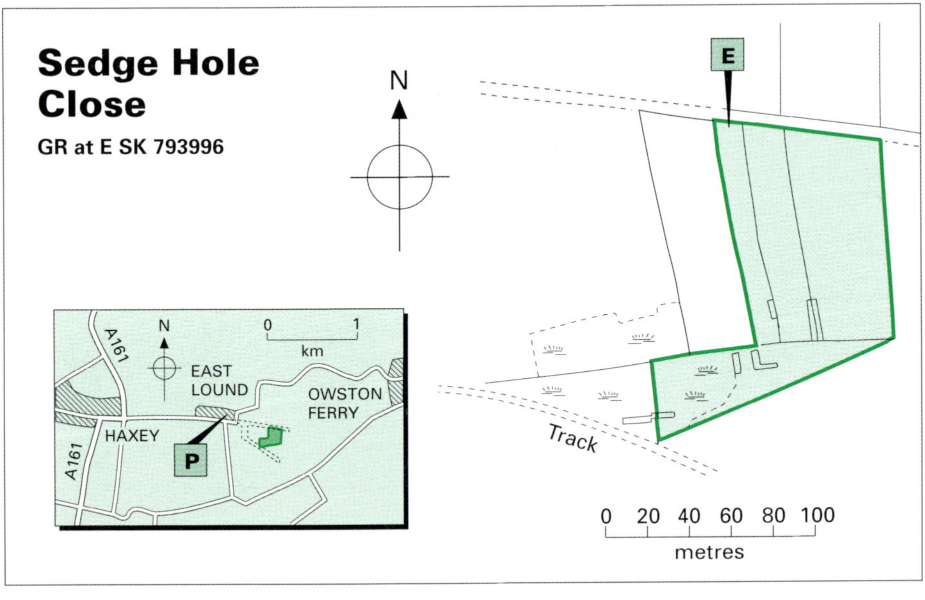

Silverines Meadows

Asterby
OS: 122   GR: TF 259787 (main entrance)   5.9 hectares (14.7 acres)
Freehold 1986
Map ref: 78

Location and Access

The meadows are situated on the south side of the village of Goulceby (although in Asterby parish). The main access is on the north side of the Goulceby–Scamblesby road from which a gate leads into the reserve at grid reference TF 259787. Cars may be parked on the roadside. It is also possible to enter the reserve at the southern end of Goulceby village via a rough track, though as this can be very wet and muddy after rain it is advisable to use the main access. Visitors are requested to keep to the edges of the reserve and to any mown paths, especially before haytime.

Description and Management

From the Red Hill reserve high on the chalk escarpment to the east you can look down towards Silverines Meadows set amid the small hedged fields that still survive on the Spilsby Sandstone soils around Goulceby. The flora includes cowslip, great burnet and devil's-bit scabious and some of the other plants found on clayland meadows, but, unusally, it also has a great abundance of the silvery-white meadow saxifrage (could this be the origin of the picturesque name Silverines?), a localised plant of sandy soils. An additional attraction is a small marsh traversed by a stream where there are southern marsh-orchid, common spotted-orchid, kingcup, ragged-robin and water avens. Snipe frequent the marsh, and the large hedges support whitethroat, lesser whitethroat and other hedgerow birds.

The meadows are sometimes cut for hay with aftermath grazing, but in most years they are let for summer grazing only.

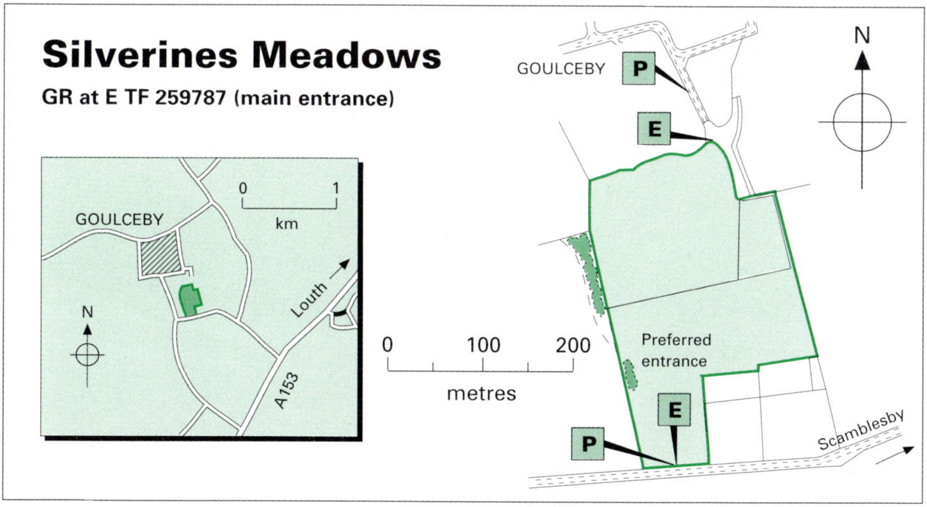

Skendleby Psalter Banks

Claxby-by-Alford
OS: 122 GR: TF 442717 1 hectare (2.4 acres)
Management agreement 1978
Map ref: 48

S ✿ SSSI

Location and Access

The reserve lies on the south side of the Skendleby Psalter valley on the Claxby–Ulceby road, about 1 km (0.6 miles) west of Claxby. The slopes may be viewed in places from the road but there is no general access for visitors. Please contact the Trust's headquarters in Horncastle for special permits or group visits. For a location plan with adjoining reserves see page 51.

Description and Management

The Skendleby Psalter valley is a dry, drift-free valley cut through the White and Red Chalk into the soft Carstone by meltwaters during the last stage of glaciation. The partially terraced chalk banks of the reserve carry a rich assemblage of chalk-loving flowers. In spring there is a spectacular display of cowslips, followed later in the year by common spotted-, twayblade and pyramidal orchids, together with quaking-grass, bird's-foot trefoil, salad burnet, burnet-saxifrage, marjoram, harebell, lady's bedstraw, field scabious and greater knapweed. Although much of the sward is dominated by upright brome, it is floristically diverse, with glaucous sedge, fairy flax, hoary plantain, rough hawkbit and, notably, adder's-tongue. Moss is prolific on the north-facing slopes, which are criss-crossed by badger paths.

Before 1971 the grassland was threatened by scrub encroachment and invasion by coarse species because of the cessation of grazing. Adjoining land, previously permanent grassland, had been turned over to arable land, leaving the steep slopes unmanaged. The management agreement provides for the Trust to arrange grazing with sheep and for continued control of invading ash and hawthorn scrub in order to maintain the floristic diversity.

Cowslips

Snipe Dales

Winceby and Lusby
OS: 122 GR: TF 319683 (Nature Reserve) and TF 330682 (Country Park)
Nature Reserve: 53.9 hectares (133 acres) 1974
Country Park: 36.4 hectares (90 acres) 1987
Map ref: 49

The Nature Reserve and Country Park are owned by Lincolnshire County Council and are managed jointly by the Council and the Trust. A full-time Warden is employed by the Trust. The reserve is leased to the Trust and is declared as a statutory nature reserve. In 1992 the Trust purchased a further 5 hectares (12.3 acres) on the eastern side of the Country Park.

Location and Access

The Country Park car park is well signposted from the A158 Skegness–Lincoln road and from the B1195 Horncastle–Spilsby road. There is also a car park for the Nature Reserve close to the hamlet of Winceby. The car parks are approximately 1.6 km (1 mile) apart and both have interpretive boards with maps. The Country Park car park is the larger of the two, and has picnic tables, an emergency telephone and public toilets. It also has toilets for visitors with disabilities and a well-graded, easy access path to central ponds and a bird hide. There is open public access, but visitors are requested to observe similar conditions to those of our members' permit and to follow the waymarked footpaths. A free colour leaflet about Snipe Dales is available from the Trust's headquarters in Horncastle or from the Warden. Dogs are permitted in the Country Park but not in the Nature Reserve.

Description and Management

Snipe Dales is situated on the southern edge of the Wolds and consists of steep-sided valleys fretted by streams, which have cut through the soft Spilsby Sandstone into the Kimmeridge Clay below. At the junction of the sandstone and clay there is a spring line giving rise to wet flushes and small streams. Snipe Dales is one of the few substantial wet-valley systems in Lincolnshire that still survive largely unspoilt. There are waymarked walks of varying length that may be followed starting from either of the car parks.

The Country Park

Snipe Dales Country Park offers a variety of attractive walks through 36 hectares (90 acres) of mainly coniferous woodland. There are picnic tables in pleasant surroundings by the car park, as well as by the central ponds and in a few other secluded areas. The woodland was previously owned by the Forestry Commission and is mainly of Corsican pine, with specimens of Scots pine and European larch. Beech grows along the southern edge of the woods, with alder, hazel and willow species in the wetter areas. The wide ride that runs to the north-east corner of the park has some magnificent mature ash trees.

Many deciduous trees and shrubs have recently been planted to give more diversity and, as sections of the pines are thinned, more plantings of native trees will gradually replace them. Nine newly excavated ponds are being colonised by water plants and attract frogs, toads and dragonflies. They are also used as stopping off points for migrant waders and waterfowl. Herons frequently visit the ponds.

Birds are best appreciated during the breeding season when bird-song and activity is greatest — at other times the pine woods may seem fairly quiet. The most abundant species are chaffinch, redpoll, willow warbler and woodpigeon. In addition to the familiar blue and great tits, coal tits are abundant. Willow tits and long-tailed tits also occur. Treecreepers and goldcrests forage in the woodlands, especially in autumn and winter. Siskins have nested in the woods; when first recorded in 1985, this was only the second breeding record for Lincolnshire. These attractive little finches may be common in the woods in winter along with small numbers of bramblings, other finches and perhaps crossbills. Both tawny and barn owls can be regularly seen, while the quick dash of a sparrowhawk may reward those with sharp eyes. Listen for the raucous cries of jays and the occasional drumming of a great spotted woodpecker. Chiffchaff, whitethroat, lesser whitethroat and sedge warbler can be heard in many places in summer, though they can be difficult to observe. The woodcock, another secretive bird, is a regular winter visitor.

There are fewer flowering plants in the Country Park than in the Nature Reserve, although generally the same species are present. Ferns are prolific among the pines, and a good selection of fungi occur in autumn.

The path waymarked with red markers, known as the 'Snipe Dales Round', provides a walk of about 6.4 km (4 miles), taking in both the Country Park and the Nature Reserve.

The Nature Reserve

There are two waymarked routes starting from the Nature Reserve car park. A short walk of about 3.2 km (2 miles), known as the 'Water Holt Trail', has orange markers. There is also the 'Snipe Dales Round' of about 6.4 km (4 miles), marked in red, which takes in the whole of the Nature Reserve and the Country Park. Along the route information boards explain in greater detail some of the features to be seen.

The reserve consists of two main valleys fretted by streams. The higher slopes are covered by rough grassland, while lower down, at the junction of the sandstone and clay, the springline is indicated by tufted hair-grass, rushes and ferns. The valley bottoms and other wet areas have a dense, tall herb flora in summer, comprising meadowsweet, great willowherb and great horsetail. Drier slopes are clothed by extensive stands of bracken and rosebay willowherb. There is evidence of much more woodland in the past: oak and ash on the higher slopes where a woodland flora with bluebells and primroses survives, and alder and willow chiefly in the lateral valleys.

Oak and ash have been planted to restore woodland cover, while alder and willow have been planted in some of the wetter holts. Prior to the Trust taking over the reserve in 1974, the grassland had deteriorated due to lack of grazing. Grazing by cattle and sheep has now been resumed with the aim of restoring a varied grass and herb community.

From the Nature Reserve car park, pass through the churchyard of the former St Mary's church. Note the rookery in the sycamore and ash trees ahead. Blackcap, sedge warbler and spotted flycatcher nest here, while swallows and house

Snipe Dales

GR at E TF 319683 (Nature Reserve entrance)

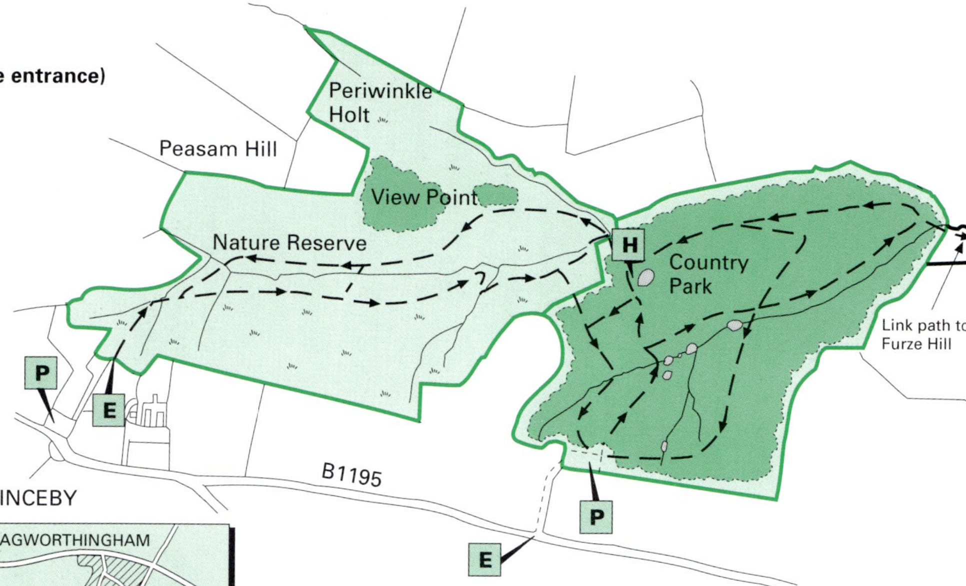

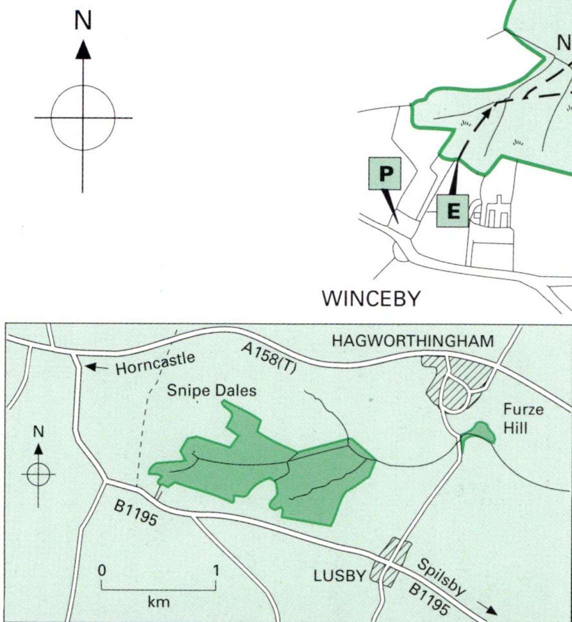

martins hawk for insects. Wild arum and violets flourish beneath the sycamore. Yellow flag, marsh-marigold, meadowsweet, lady's smock, marsh thistle and primrose bloom in the holt bottom.

An old hydraulic ram thuds regularly in the valley bottom as it pumps stream water to the farm on the hill at Winceby. The footpath follows the stony bedded stream, which is rich in aquatic invertebrates and where water voles may be seen. The streamside plants include ragged-robin, water avens, water figwort, marsh-bedstraw, water mint and water-cress. Snipe and meadow pipit are found in the grazed paddocks.

Eventually the path turns left up the hillside, the home of grasshopper warbler, whitethroat, willow warbler, linnet, yellowhammer and reed bunting.

A short detour path leads to a viewpoint from which the Wolds can be glimpsed through oak trees planted in the 1970s. From here you can see much of the reserve and the changing vegetation at the springline. Barn owl, short-eared owl and kestrel regularly hunt here, and green woodpecker may be heard.

Continue past a small plantation and gorse thicket to the main stream, where a path to the Country Park ascends the slope to the left (the red route). The orange path (Water Holt Trail) passes through a wet area with betony, tormentil, bird's-foot trefoil, alder and willow. Eighteen species of butterfly have been recorded, some of which may be seen on the brambles along the next dry section of the route, where there are also common spotted-orchid, meadow saxifrage and great horsetail. The path rejoins the entrance path at the footbridge.

Link Path to Furze Hill and Sow Dale: 8 km (5 miles)

The path is accessible from both car parks and is waymarked with white arrows along its entire length. It continues alongside the alder-lined brook, passing through the extension at the eastern end of the Country Park, and leads to the ford at Hagworthingham. Cross the road and find the path, which continues through Furze Hill. After crossing the bridge follow alongside the brook to the eastern end where a public footpath re-crosses the brook. This path follows the brook for a kilometre or so before branching right along a hedgerow and climbing up the open grassland of the hillside. A headland path then crosses the ridge of land, which carries the B1195 road. Cross the road to a field edge path, which leads to the head of Upper Sow Dale (see page 177). The path drops steeply into the alder/birch wood before emerging on the open meadows of the valley bottom, which are grazed in summer and autumn by cattle.

Lady's smock

Sotby Meadows

Sotby
OS: 122 GR: TF 205778 6.1 hectares (15.2 acres)
Part freehold, part leasehold 1979
Map ref: 50

SSSI

Location and Access

The reserve lies about 1 km (0.6 miles) south of Sotby. Passing southwards through the village take the right fork towards Wragby. A green lane will be found on the left, where cars may be parked. The reserve entrance is reached on foot: walk down the green lane and the entrance gate will be seen on the left. Visitors are requested to keep to the field footpath and avoid trampling, especially before hay time.

Description and Management

These small meadows with the adjoining green lane and large old hedges form a landscape and wildlife feature of special interest, representing a type of countryside that has now almost vanished in Lincolnshire. The flora includes adder's-tongue, moonwort, dropwort, great burnet, yellow rattle, twayblade, and pyramidal and green-winged orchids. Around 200 species have been recorded, including 28 grasses. The hedges are fine examples of old species-rich hedgerows. The green lane hedge adjoining the reserve has oak, ash, hawthorn, purging buckthorn, blackthorn, crab apple, midland hawthorn and field maple. Seventeen breeding species of birds have been recorded in the hedgerows.

Management consists of taking a late summer hay-crop and grazing the aftermath during autumn.

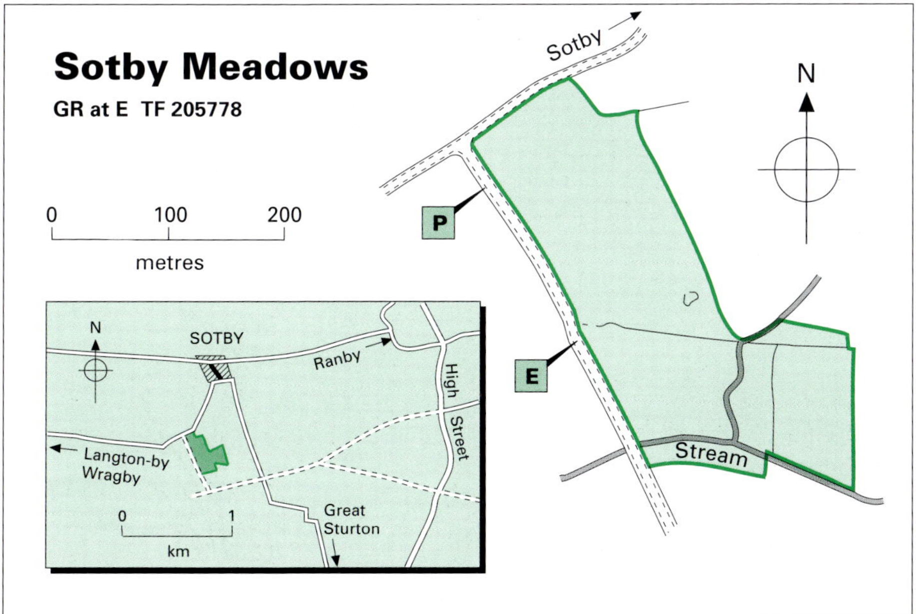

South Witham Verges

South Witham
OS: 130 GR: SK 938193 6 hectares (14.8 acres)
Management agreement 1991
Map ref: 96

Location and Access

The reserve lies due east of the village of South Witham at the intersection of the road to Castle Bytham and the A1. The reserve includes the slip roads to and from the A1 and a small paddock. The verges and embankments can only be approached from the road and great care is needed because of the fast-moving traffic along the road that divides the banks. For this reason it is not suitable for parties to visit. There are two private field access points at each end of the reserve where there is parking space for two cars but allowance should be made at all times for access by farmers with big machines or tractors.

Description and Management

When the A1 was upgraded to a dual carriageway in 1970, the old South Witham–Castle Bytham road was realigned along a disused railway line. The cutting was widened to allow for two lanes of road traffic; the north bank is part of the original cutting and the south a newly created one. The reserve includes grassy banks and verges and large areas of scrub and hedgerows.

There is a rich limestone flora on the grassland consisting of such specialised flowering plants as kidney vetch, eyebright, yellow rattle, oxeye daisy, bee orchid, common knapweed, blue fleabane and fairy flax. There are also extensive areas of early dog-violet, cowslip, wild carrot, wild parsnip, bird's-foot trefoil, black medick, melilot, beaked hawk's-beard, red bartsia and glaucous sedge, and many grasses. Twenty-eight species of butterfly have been recorded, including five kinds of skipper, green hairstreak, clouded yellow, small heath and dark green fritillary. Mammal records include fallow and muntjac deer, badger, fox, stoat, weasel, pygmy shrew and harvest mouse.

The scrub on the higher slopes consists of hawthorn, blackthorn, hazel, sallow, dogwood, gorse and rose. A wild pear can also be seen, only the second to be recorded in the county. The scrub forms what is probably an optimal habitat for songbirds, with blackbird, song thrush, robin, dunnock, wren, long-tailed tit, greenfinch, bullfinch and yellowhammer being regular breeding species. In 1995, 31 pairs of willow warblers, eight common whitethroats, seven garden warblers, five lesser whitethroats and four blackcaps were recorded. Altogether 100 bird species have been recorded of which 34 have bred.

In the ditches grow common spotted-orchid, meadowsweet, marsh thistle, great willowherb, common fleabane, water-cress, great reedmace, yellow iris and false fox-sedge.

The floristic diversity is maintained by an annual cut of the grass and selected scrub, particularly thorn, rose and ash, which is done after the flowering season in September or October. This has encouraged an increase in early dog-violet, kidney vetch and oxeye daisy, and orchids have appeared in new areas.

South Witham Verges

GR at SK 938193

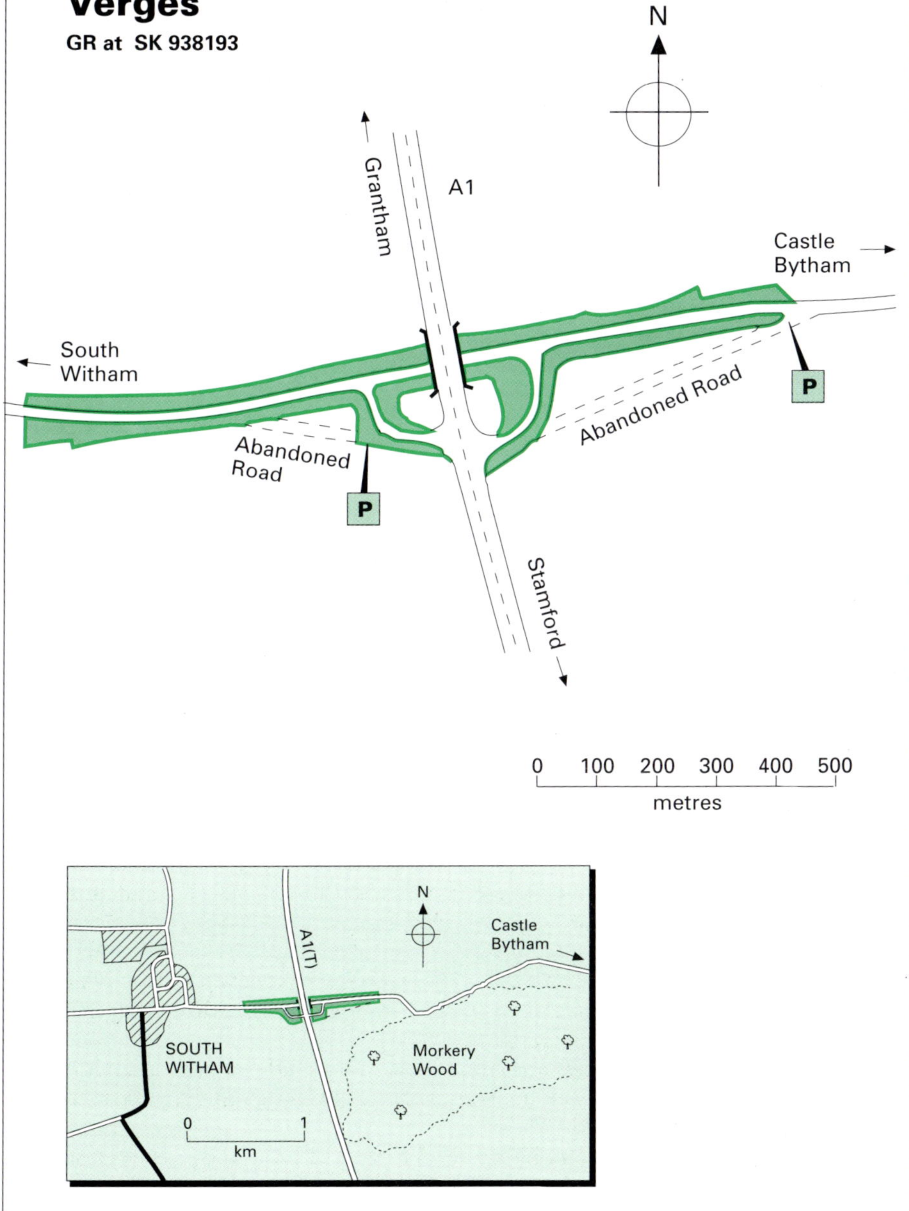

Sow Dale

Old Bolingbroke and Mavis Enderby
OS: 122 GR: TF 349666 (Upper Sow Dale) TF 351655 (Lower Sow Dale)
34 hectares (84 acres)
Freehold 1986
Map ref: 74

Location and Access

The access to Lower Sow Dale is close to Old Bolingbroke, from the Spilsby road at TF 352653, while Upper Sow Dale is reached from the B1195 Horncastle–Spilsby road approximately mid-way between Lusby and Mavis Enderby at TF 349672, where there is roadside verge parking. A track to the south side of the B1195 road, fenced from the adjoining field, leads to the reserve where an attractive path with steps has been constructed through the ash wood into the bottom of the valley and thence into the pasture below. The track can become somewhat soft after rain and the climb back up the steps is fairly steep. There is a link path from Upper Sow Dale to Furze Hill and Snipe Dales (see pages 83 and 171).

Description and Management

Sow Dale is the most extensive of the dales that fret the Spilsby Sandstone country of the southern Wolds. For centuries it has formed an important access to the hills from the fenland country to the south; the famous prehistoric site of Hall Hill overlooks the entrance to it, and by the Middle Ages it was guarded by John of Gaunt's great castle at Old Bolingbroke. Until the early 1980s the Dale had survived in traditional usage of meadow, pasture and woodland, but recently more intensive farming has converted large areas of it to arable cropping, with consequent loss of landscape and wildlife interest.

The Trust's reserve consists of a meadow and woodland in the upper part of the Dale and a block of some 23.5 hectares (58 acres) of pasture in lower Sow Dale immediately adjacent to Old Bolingbroke. There are some fine old hedges, and willow and alder carr alongside the stream. Common spotted-orchid, ragged-robin and fleabane are among the flowers of the grassland; kingcups make a bold splash of colour in April in the spring-line flushes on the valley sides where the sandstone meets the Kimmeridge Clay below it. In the carr the stands of giant horsetail resemble a coal measure forest and the branching marsh thistles grow to 6 feet or more. Both species of golden-saxifrage (alternate-leaved and opposite-leaved) grow in shady places by the stream, and in the open there are water forget-me-not and brooklime. Badgers are regular in the Dale and there is a varied bird fauna. Kestrel, sparrowhawk, barn and tawny owls, snipe, and green and great spotted woodpeckers are all resident. Garden, grasshopper and sedge warblers, blackcap, whitethroat and lesser whitethroat, and spotted flycatcher are summer visitors. The common grassland butterflies are usually abundant.

The grassland is grazed by sheep and/or cattle. Invasion by rushes has to be controlled. A woodland management programme is designed to maintain the ash wood and the alder/willow carr.

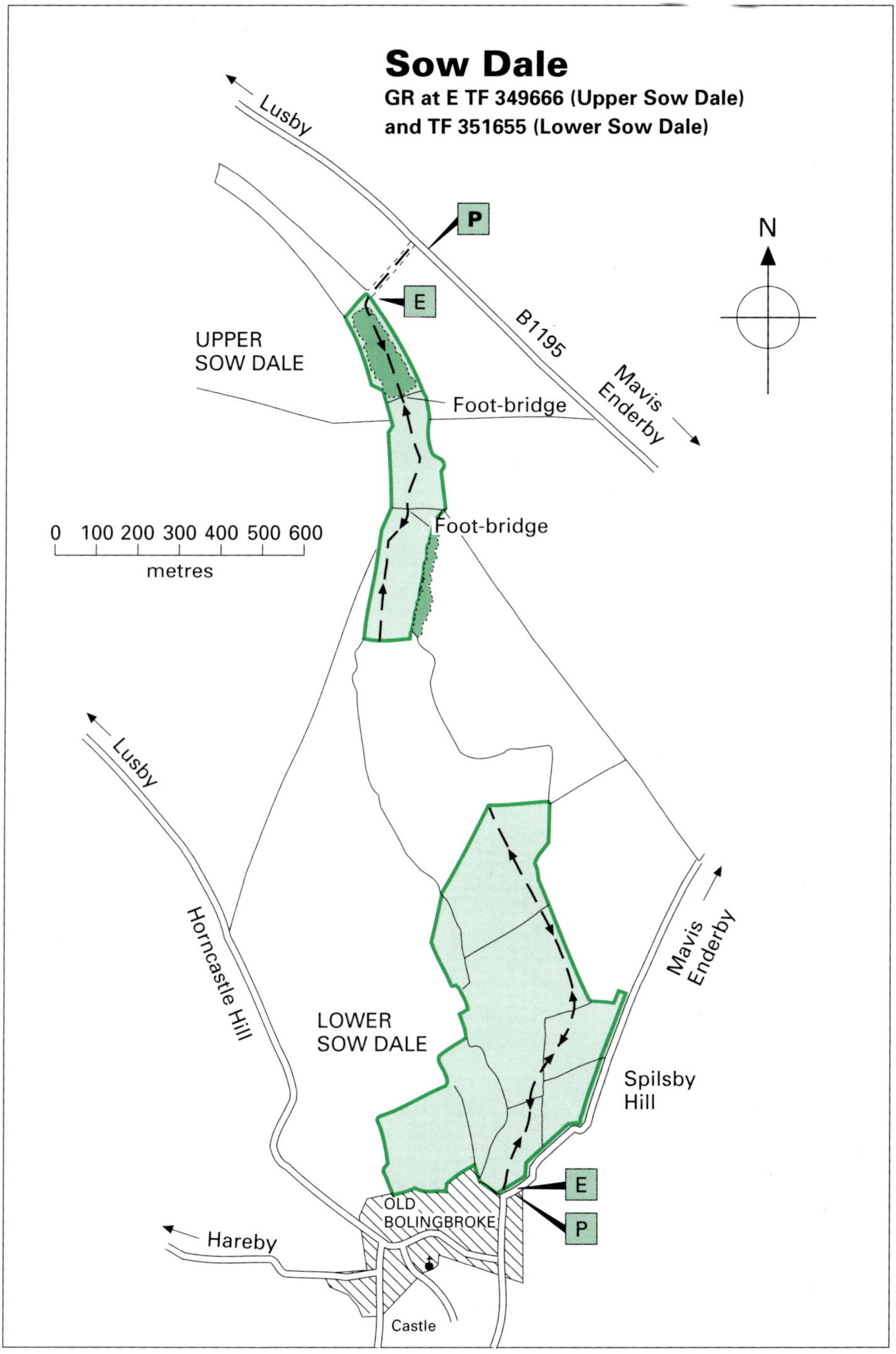
Sow Dale
GR at E TF 349666 (Upper Sow Dale)
and TF 351655 (Lower Sow Dale)
Lusby
P
E
N
UPPER
SOW DALE
B1195
Mavis
Enderby
Foot-bridge
Foot-bridge
0 100 200 300 400 500 600
metres
Lusby
Horncastle Hill
Mavis
Enderby
LOWER
SOW DALE
Spilsby
Hill
E
P
OLD
BOLINGBROKE
Hareby
Castle

Stanton's Pit

Witham-on-the-Hill
OS: 130   GR: TF 032174   8 hectares (19.8 acres)
Freehold 1996. Gift from Mr RA Stanton
Map ref: 103

Location and Access

The reserve is situated on the south side of the minor road between Little Bytham and Witham-on-the-Hill. Visitors are advised to use the entrance to Bullimores weigh-bridge and to park in the reserve car park (GR: TF 034172) to the left of the gated reserve entrance. The main access routes around the reserve will be suitable for wheelchair use once they have bedded down. There is no access along the margin near the road.

Description and Management

The area was formerly a sand pit and is a gift from the former owner for use as a nature reserve. The primary interest in its present form is ornithological. Recent records show that over 50 species of birds have visited the site, of which 19 species have bred. There is some evidence to suggest that the reserve lies on a migratory route from the Wash inland towards Rutland Water. Among breeding birds are little grebe, little ringed plover, sand martin, turtle dove and lesser whitethroat. In autumn migrant waders include little stint, ruff, spotted redshank, greenshank, and common, curlew, green and wood sandpipers.

In addition to open water there are extensive areas of grassland, including some of the original field. There are wet and poorly drained areas in the eastern section. For those visitors wishing to follow the track around to the plateau, extensive views from the top of the large mound can be had.

There are plans for major tree and shrub planting schemes to add habitat diversity and to provide cover to the hides. In time, marginal vegetation will develop around the lake and there may be scope for islands, tern rafts and shallow scrapes, although with two land drains entering the lake, water-level management may prove difficult.

Tufted duck and pochard

Stanton's Pit

GR at E TF 032174

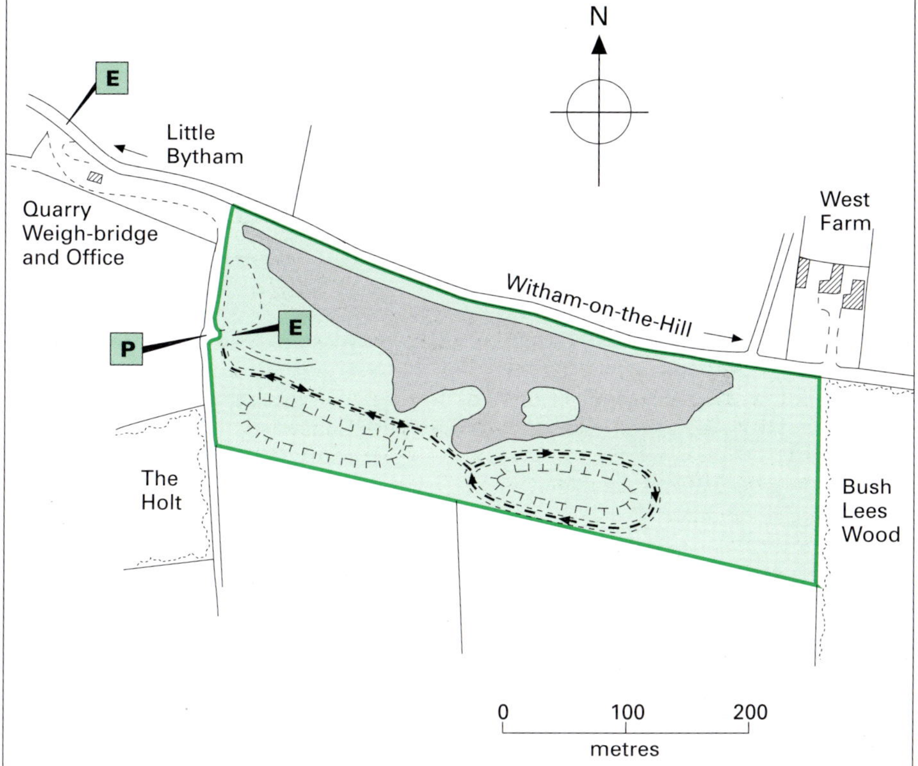

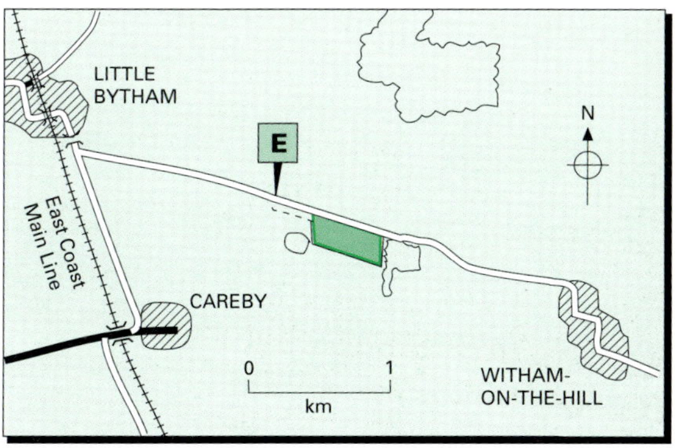

Surfleet Lows

Surfleet
OS: 131 GR: TF 251287 3.4 hectares (8.4 acres)
Leasehold 1968
Map ref: 51

SSSI

Location and Access

The reserve is situated on the east side of the old A16 Boston–Spalding road, about 300 m north of Surfleet church. There is parking space on the opposite side of the road between the village and the reserve. Access is via the stile and foot-bridge into the woodland area.

Description and Management

Most of the reserve consists of old grassland and is subject to winter flooding. The low-lying area is the course of an old creek, which joined the Old Beck in the field to the south of the reserve. Although there is some doubt about the past history of this area, it seems that it may have been saltmarsh within the tidal limits of the River Glen in the Middle Ages. The level of salinity in parts of the meadow is still remarkably high, supporting salt and brackish marsh species, such as sea-milkwort, distant sedge, glaucous rush and golden dock. In addition, there is a rich flora of grasses and sedges, and five species of buttercup, including the localised hairy buttercup.

The small area of damp woodland contains some old willows, and has been partially replanted with sallow, crack willow, ash, alder and oak to replace diseased elms. Nearby are small pools surrounded by a reedbed.

More than 50 species of birds have been recorded. Snipe occur in winter, when redshank, mallard, teal and wigeon are also attracted by flood water. There is a small colony of reed warblers in summer.

The meadow is grazed by cattle from May to November to maintain its floristic interest. Along the roadside the hawthorn hedge was traditionally managed by cutting at 10-year intervals to provide river-training faggots, but more recently it has been laid to improve its function as a stock-proof boundary.

Hairy buttercup

Surfleet Lows

GR at E TF 251287

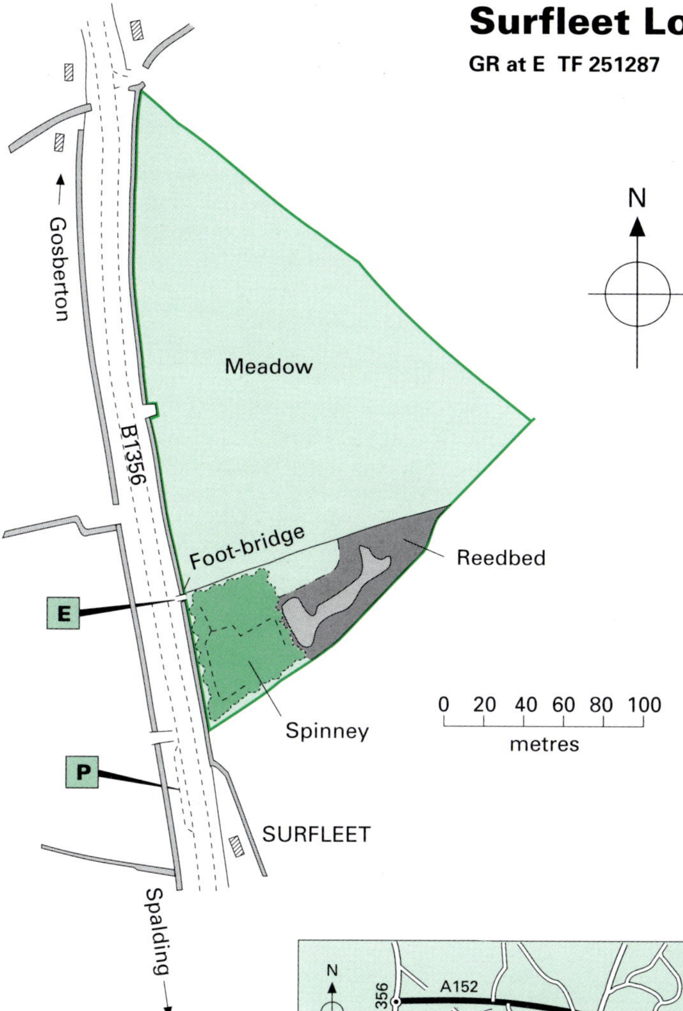

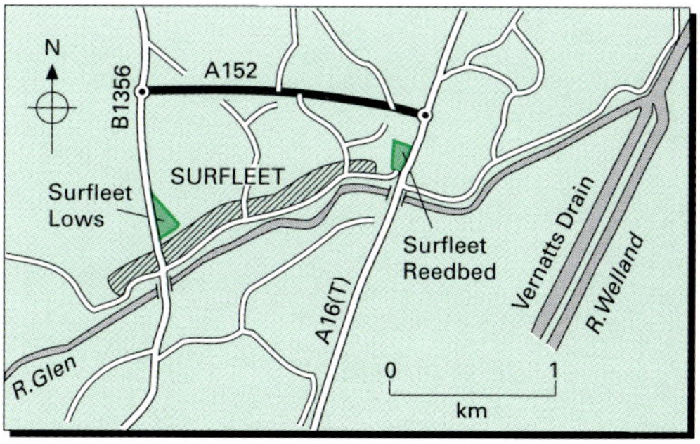

Surfleet Reedbed

Surfleet
OS: 131   GR: TF 266289   1.4 hectares (3.3 acres)
Leasehold 1970
Map ref: 52

S

Location and Access

Situated 1.5 km (0.9 miles) east of Surfleet church, the nature reserve can be viewed from a footpath leading north from the gateway adjacent to the junction with the new A16. Cars should not be parked where they will obstruct the field gateway. Visitors are warned that they must not attempt to enter the reedbed because of the unstable nature of the reed litter floating on waterlogged organic mud. For a location plan with a nearby reserve see page 181.

Description and Management

The reserve, formerly a railway borrow-pit, contains one of the largest single areas of common reed *Phragmites* in the Spalding district and provides a nesting habitat for reed and sedge warblers. In autumn it is used as a roost by swallows, martins and wagtails. Woodcock, water rail and long-eared owl have occurred in the sallow scrub. The scrub, or carr, is slowly encroaching into the reedbed. The reedbed, which supports common skullcap and lesser reedmace, floats on a foul-smelling black ooze that is frequently flooded above the litter layer after heavy rain. In the centre of the reedbed there is a pond, the edges of which are extremely dangerous. The construction in 1994/95 of the new A16 (Spalding to Boston) road has added extra land to the reserve adjacent to the new road.

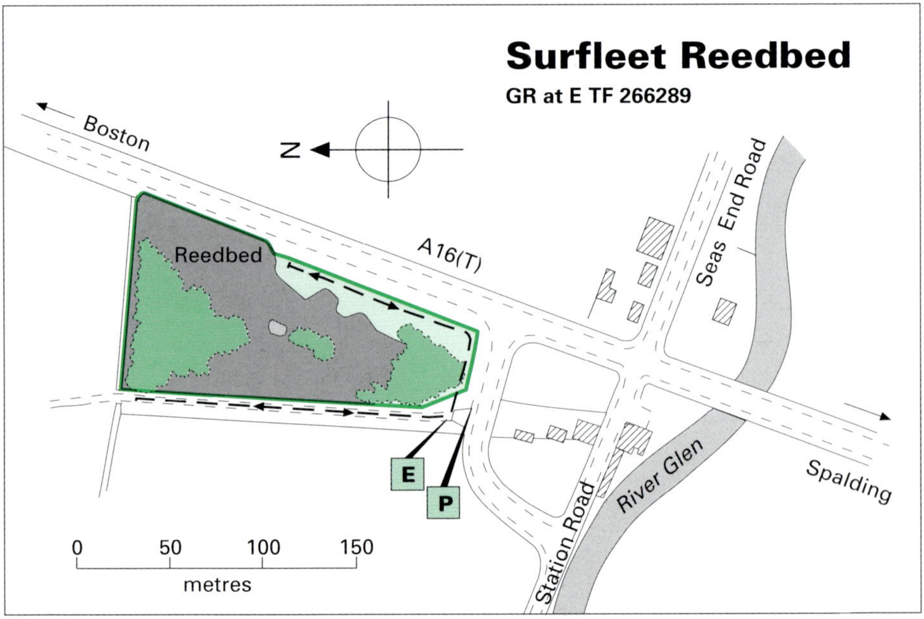

Swaby Valley

Swaby
OS: 122   GR: TF 390776   3.6 hectares (9 acres)
Part management agreement, part freehold 1979
Map ref: 53

SSSI

Location and Access

The reserve lies in the valley of a tributary of the Calceby Beck immediately to the east of Swaby village. A public footpath runs through the valley from the end of Valley Lane. Please do not cross the stream: access to the land to the south is by special permit only. Cars should be parked in the village as there is no parking space close to the valley.

Description and Management

This is a fine example of a glacial overflow valley with chalk grassland on the terraces and slopes and a rich calcareous marsh on the valley floor. Chalk plants include wild thyme, small scabious, salad burnet, burnet-saxifrage, pyramidal and bee orchids. Sixteen species of butterfly have been recorded, including common blue and small copper. There is a pleasant chalk stream in the valley with an interesting invertebrate fauna and marginal vegetation.

Management consists of maintaining a proper grazing regime by agreement with the owners.

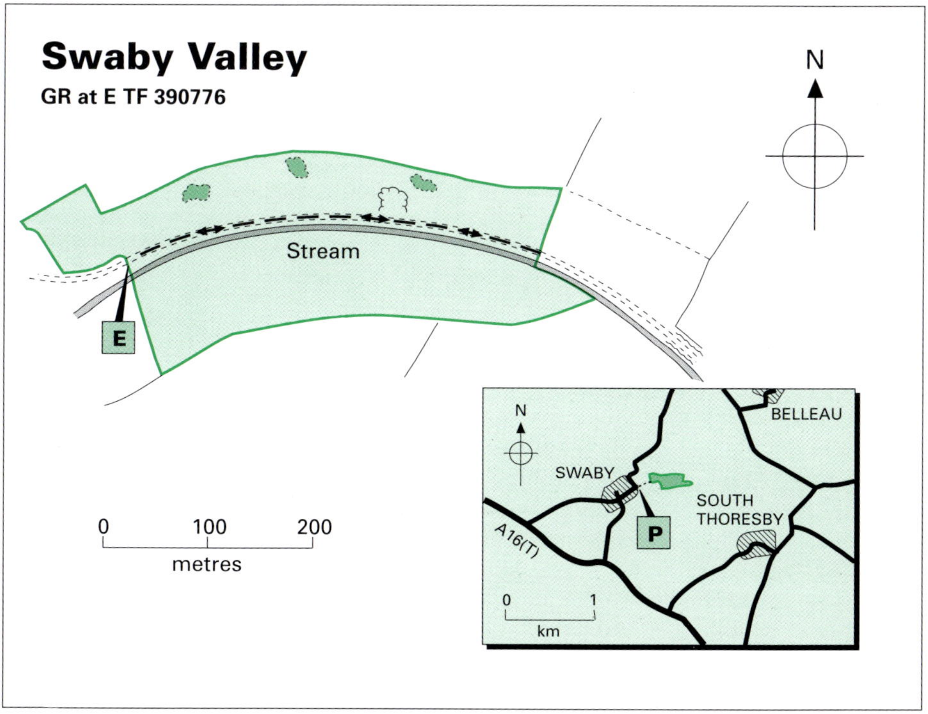

Tetney Blow Wells

Tetney
OS: 113 GR: TA 320007 14.9 hectares (36.8 acres)
Management agreement 1972
Map ref: 54

SSSI

Location and Access

The reserve lies south-east of Tetney church. The entrance, which also leads to the Anglian Water Authority pumping station, is approximately 150 m east of the church along Church Lane. Cars may be parked along Church Lane, near the church. Cross the stile beside the gate marked 'No Unauthorised Access' and walk up the main drive towards the pumping station. Visitors are requested to keep to the defined paths or waymarked route, which avoids dangerous boggy areas and deep water.

Description and Management

The reserve consists of four large blow wells with associated habitats, which include damp woodland, meadow and old water-cress beds that have been colonised by marsh vegetation and reedbeds. The blow wells are the result of artesian springs where water under pressure escapes from the chalk through structural weaknesses in the overlying boulder clay to reach the surface. The wells, which are reed-fringed and have prolific growths of starwort and Canadian waterweed, are used by Anglian Water as a domestic water supply source. The area was used as a water-cress farm from 1948 to 1961 and the remains of concrete channels and pathways, which served four water-cress beds, may still be seen. These beds are now dominated by reeds and great willowherb, and one in the south-west is regularly cut and raked in late summer to maintain its flora. The grassland areas provide a fine display of cowslips in spring and are mown for hay in July by a local farmer.

Waymarked Route 1.2 km (0.7 miles)

Follow the waymarked route from the Anglian Water Authority pumping station, passing the nearby blow well and crossing the meadow as you walk towards the west. You will soon pass a second blow well where its overflow channel flows under the path on its way to the river.

Looking north you will see Tetney church, built in 1363, with its fine tower of Yorkshire limestone added 100 years later. The church was considerably extended in the 19th century.

As you continue you will see Scots pine in the corner to your right interspersed with sallow and hawthorn. Turn left along the bank between the blow well and the boundary dyke. Look for water-starwort and spikes of mare's tail in the water. You may also see coot or moorhen on the water, and sometimes mallard and tufted duck. Sedge warblers are often to be heard and seen in the surrounding reeds and bushes. Walk along the concrete channel and follow this south, parallel to the disused water-cress beds, now overgrown with reed, on your left. Near to the southern boundary the route turns left across the reedbed, where you may see great willowherb, reedmace, figwort, meadowsweet, hemp-agrimony, gipsywort and marsh-orchid.

Tetney Blow Wells

GR at E TA 320007

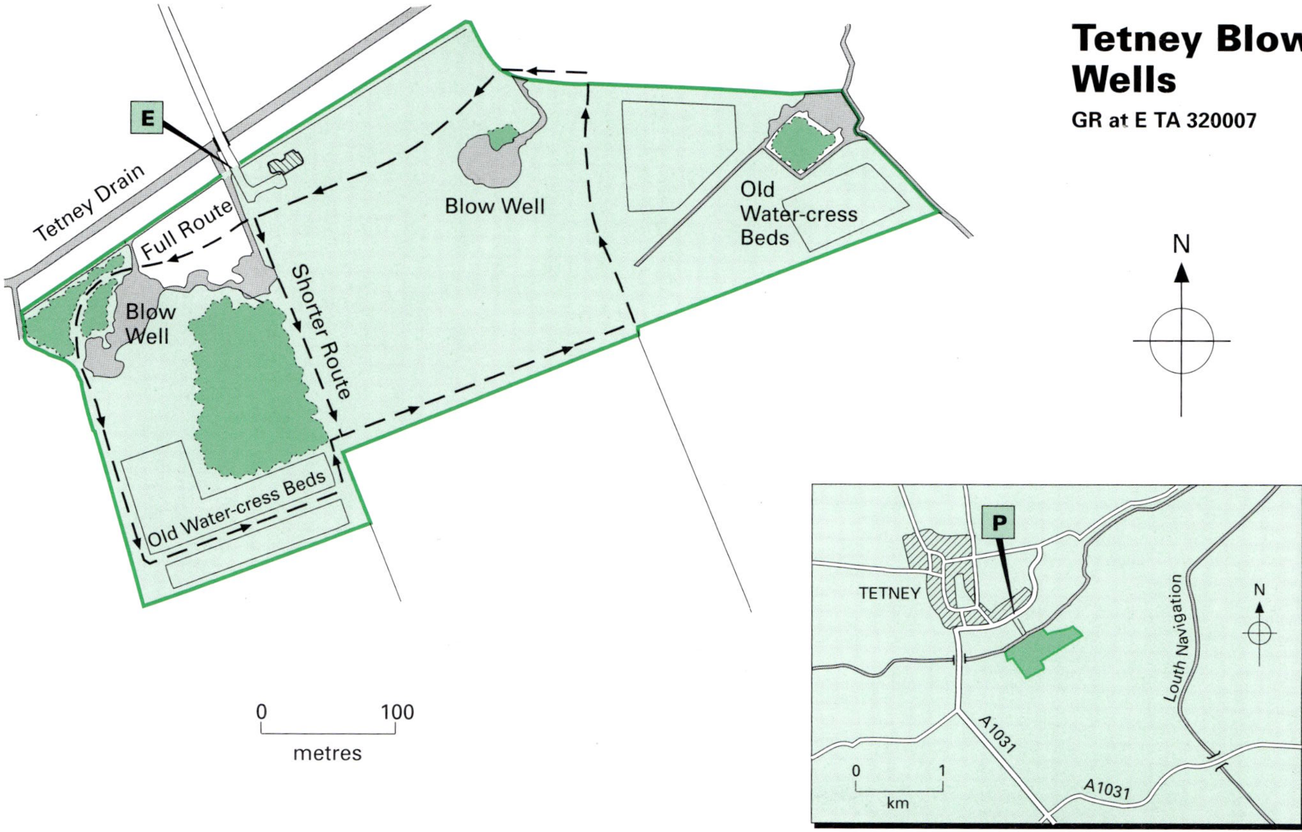

On leaving the reedbed, turn left into the corner of the copse, where there are planted willows, alder, poplar and ash. Follow the path along the southern boundary dyke, in which floating pondweed, water-plantain and water-crowfoot grow. Look for the purple flowers of lesser knapweed in the meadow. In spring you may see cowslips in the meadow beyond, and in late summer knapweed can be seen near the dyke. Turn left after the hawthorn hedge and follow the grassy track beside the poplars, and then cross the meadow towards the gate. You will pass more disused water-cress beds on the right, now covered with scrub.

Through the gate, turn sharply left along a public footpath by the dyke-side and cross a foot-bridge. In summer, if you wish to see water-violet in flower, you may visit a blow well at the north-east corner of the reserve by following the public footpath to the right after the gate and continuing over the stile. Return along the path across the meadow to your starting point by the pumping station. If you wish to take a closer look at this blow well, turn left immediately after the foot-bridge along a short path, which leads to the water overflow channel.

Tetney Blow Wells

The Chasm & Northorpe Slipe

Bourne and Thurlby
OS: 130 GR: TF 151185 & TF 129170 13.9 hectares (34.5 acres)
Management agreement 1996
Map ref: 80

Location and Access

Access to The Chasm is from the road between Bourne and Tongue End. The reserve is reached on foot via a roadside gate on the north bank of the River Glen at Tongue End Bridge. There is space for cars on the roadside verge. The reserve was extended in 1996 to include the Northorpe Slipe and river bank linking up with Thurlby Fen Slipe. Access to Northorpe Slipe is from the end of Long Drove, adjacent to the east end of Thurlby Fen Slipe, where cars may be parked.

Description and Management

The Chasm was once a land drainage pumped-storage reservoir. It has been disused for many years, and part of the impounding banks have been removed. The remaining wide ditch provides the main interest for the site, and the marginal vegetation includes plants that are scarce in the Fens, such as flowering-rush and greater water-parsnip. Other plants to be found include lesser water-parsnip, broad leaved pondweed, marsh pennywort and purple loosestrife. Northorpe Slipe consists of a broad strip of grassland, on peat, in which borrow-pits were excavated after the 1947 floods to provide material for re-building the river banks. Parts of these pits contain good stands of reed, while other sections are dominated by bulrush (reedmace) and sedges. Small areas of open water contain broad-leaved pondweed and water violet. It is hoped to develop the wetland habitat further by extending the area of reed and by creating more open water. The whole area is of interest for dragonflies, such as migrant and southern hawker, ruddy darter, four-spotted chaser and emerald damselfly. Nesting birds include sedge and reed warblers, and mute swan. Water rail occurs and the site is a favoured hunting area for barn owl.

The site is grazed by sheep, which control the growth of tall herbs and scrub on the grassland and river bank.

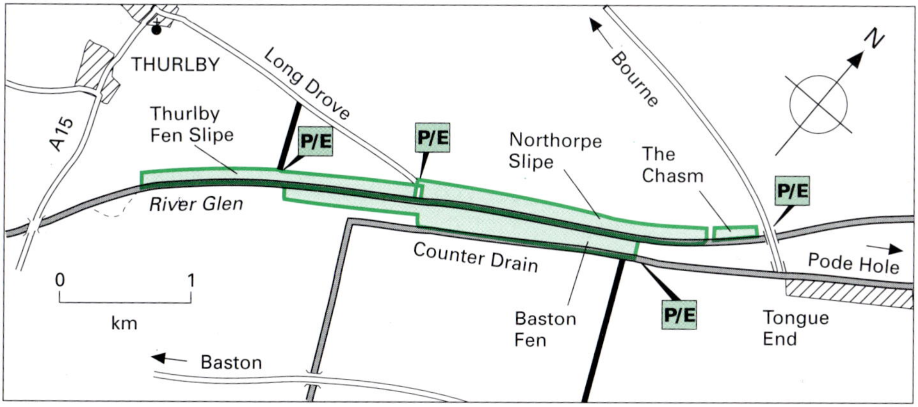

The Shrubberies

Long Sutton
OS: 131 GR: TF 425234 4.5 hectares (11.2 acres)
Leasehold 1971
Map ref: 55

Location and Access

Situated at the western end of Long Sutton, the reserve lies to the south of the old A17. The entrance is via a gate from a small side road (Garnsgate Road) where cars may be parked.

Description and Management

The Shrubberies comprise old parkland and pasture of a type now rare in the Fens with fine oak and other large trees. Some 49 species of birds and 12 species of butterflies have been recorded. There is a pond with a wooded island and adjoining marshy areas with fringing alders. The grassland is grazed and sometimes cut for hay. An acre of land was planted with native trees in 1989.

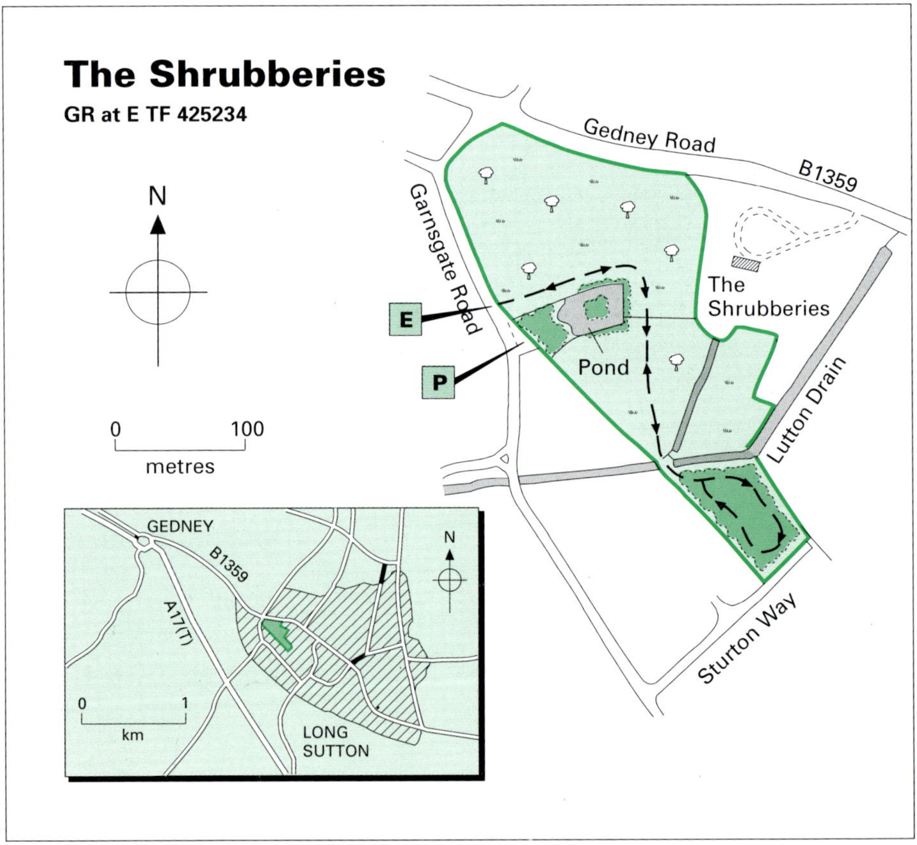

The Yews

Donington
OS: 131 GR: TF 220345 1.6 hectares (4 acres)
Management agreement 1971
Map ref: 56

S

Location and Access

The reserve is located to the west of the A152 road, some 2 km (1.25 miles) south of Donington.

Description and Management

The Yews is a compact small farmstead consisting of old farm buildings set in permanent and ley grass meadows with fine hedgerows. Old enclosure grassland and good hedges on silt are now rare in the Fens, and the site has considerable potential for conservation improvement by management. Trees and shrubs have been planted in one of the ley fields and will be developed to replace Fen woodland, lost locally in the 1940s. The remaining grassland is grazed with sheep.

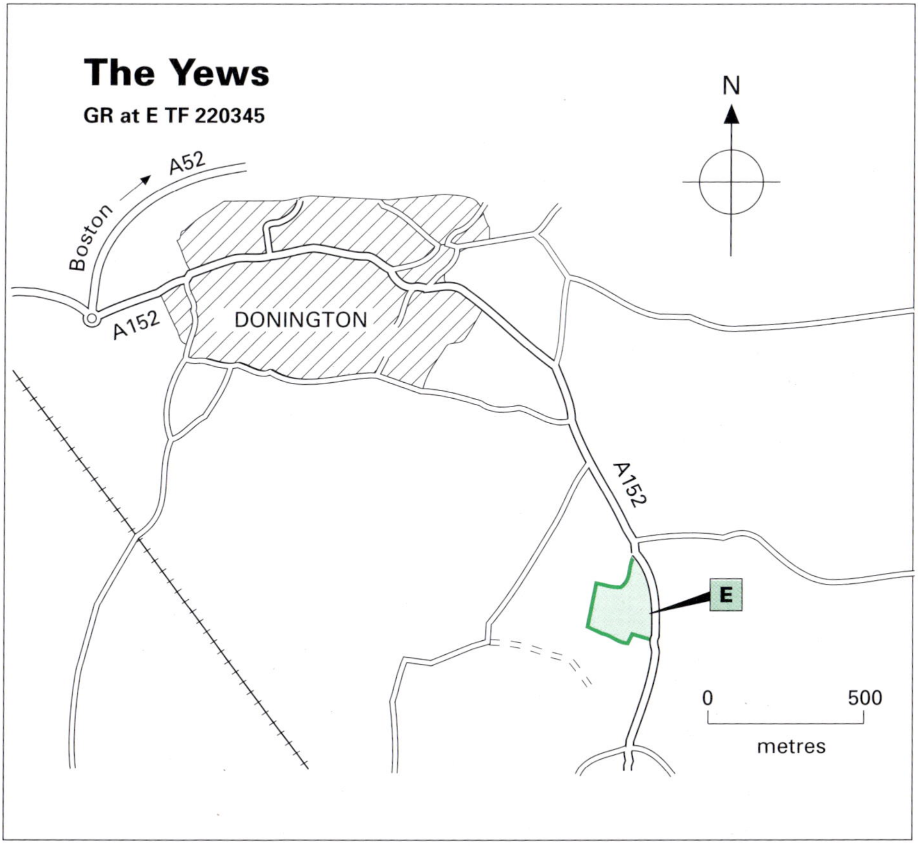

Thurlby Fen Slipe

Thurlby
OS: 130 GR: TF 119164 7.8 hectares (19.2 acres)
Freehold 1988
Map ref: 82

Location and Access

Situated on the north side of the River Glen opposite the Trust's Baston Fen nature reserve, Thurlby Fen Slipe is reached from the A15 by turning east at Thurlby crossroads. Passing Thurlby church on the right, follow the lane for 1.5 km (0.9 miles) and then turn right along Baston Edge Drove green lane to the reserve entrance where there is parking space for cars at the end of the drove road. There is a stile at each end of the reserve giving access to the river bank so that a circular route may be walked. For a location plan with adjoining reserves see page 187.

Description and Management

Slipe is the land at the foot of a river bank, and this reserve, which is about 2 km (1.2 miles) long, consists of old flooded borrow-pits with a gravelly bed, associated reedbeds, scrub and grassland and adjacent hedgerows. The grassland area to the east of the reserve was arable until 1986, and part of it has been scraped to encourage an extension of the main reedbed.

The pools have a rich flora, which includes scarce marsh and aquatic plants such as greater tussock-sedge, greater spearwort, water-violet, fen pondweed and bladderwort. Purple-loosestrife, marsh pennywort, marsh-marigold and common spotted-orchid are also among the 210 species of flowering plants that have so far been recorded on the site. Cowslips make a spectacular show in spring. Fine chub lie in the clear pools. Birds are numerous and varied: 109 species have been recorded, of which 40 have bred. The reserve is noted for its dragonflies with 15 species regularly recorded, 14 of which are known to have bred. Twenty-three species of butterfly have been recorded, most favouring the west end of the reserve.

Management has involved clearance of some of the overgrown pools, scrub control, hedge laying, path maintenance and conversion of the former arable land into meadow and wetland habitats. The water-level in the pools and marshy areas is controlled by a sluice.

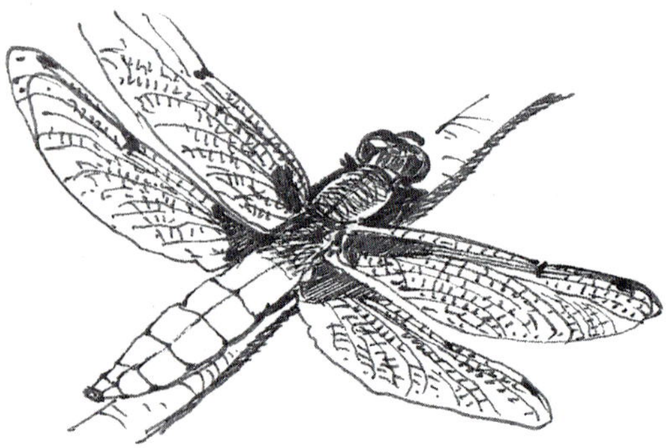

Broad-bodied chaser

Thurlby Fen Slipe

GR at E TF 119164 (Baston Edge Drove entrance)

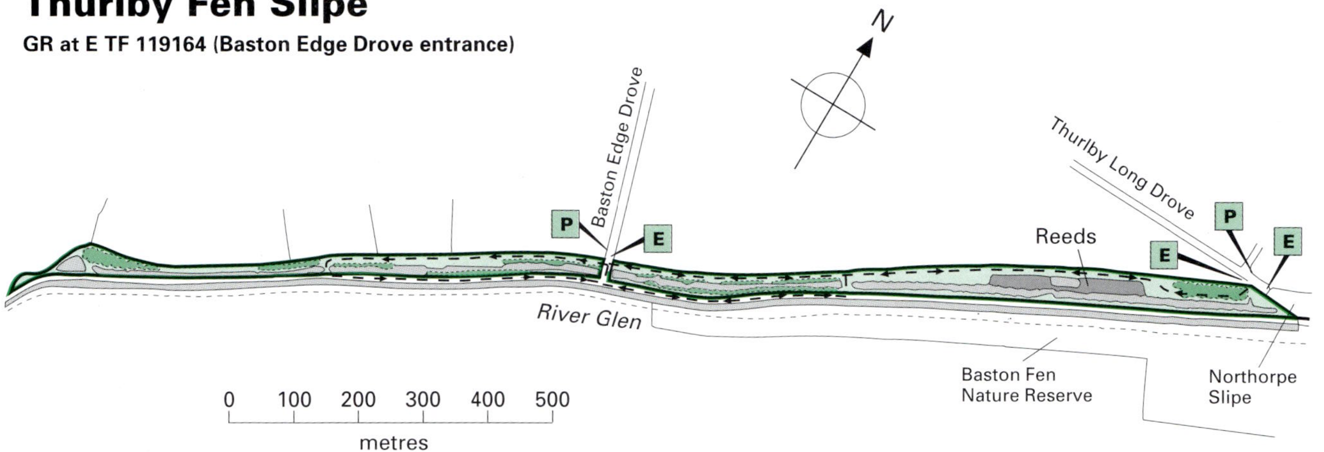

Toft Tunnel

Bourne
OS: 130 GR: TF 077190 3 hectares (7.6 acres)
Management agreement 1993
Map ref: 101

Location and Access

The reserve lies south-west of Bourne on the A6121 Stamford road, about 1 km (0.6 miles) from the A151 junction. Limited roadside parking is available by the entrance to the west cutting. Parking at weekends and evenings is available in the SKDC Lound Depot, courtesy of SKDC (please park adjacent to the east cutting and display a Trust car sticker). Access to either cutting is from the roadside. Access to the track level is by steps.

Description and Management

Part of the former Bourne–Saxby railway line linking the Midlands to the East coast, the reserve consists of two deep cuttings and the land over the 300-m-long tunnel. The track level of both cuttings can become very wet in winter, the east cutting particularly so. An all-weather raised path in the east cutting has been constructed. The mixture of scrub and open areas with species-rich grassland provides a diverse range of habitats. Whitethroat and willow warbler are regular nesting species, while in winter there are often large numbers of fieldfare and redwing. Twenty-one species of butterfly have been recorded.

Management mainly consists of maintaining areas of dense hawthorn and blackthorn scrub while restoring some areas of permanent grassland.

Toft Tunnel

Toft Tunnel

GR at E TF 077190

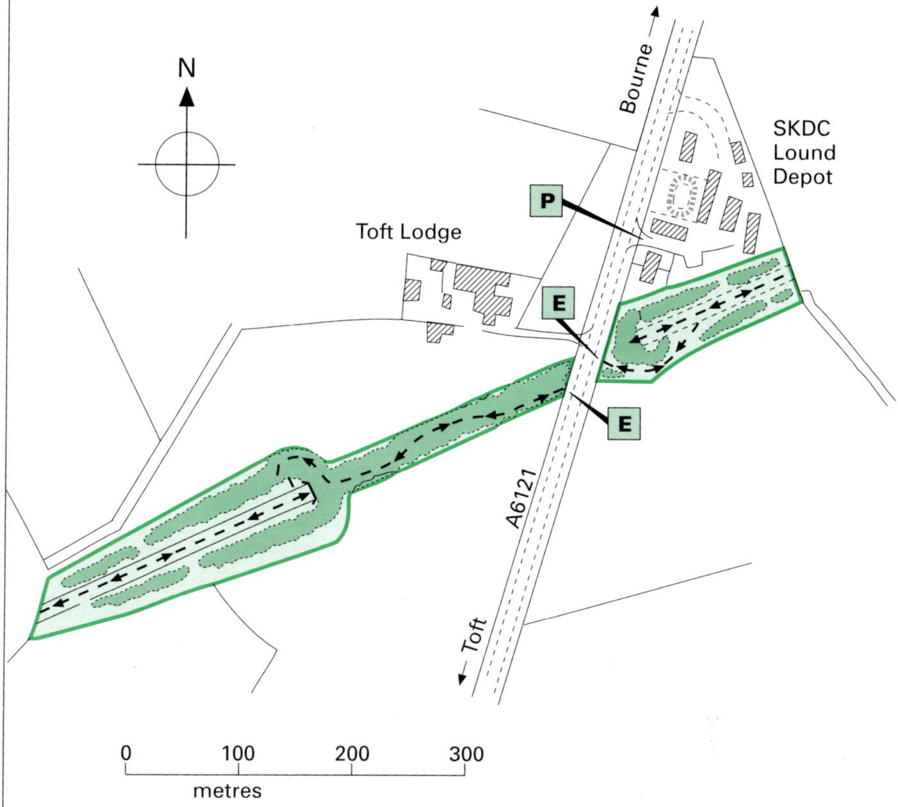

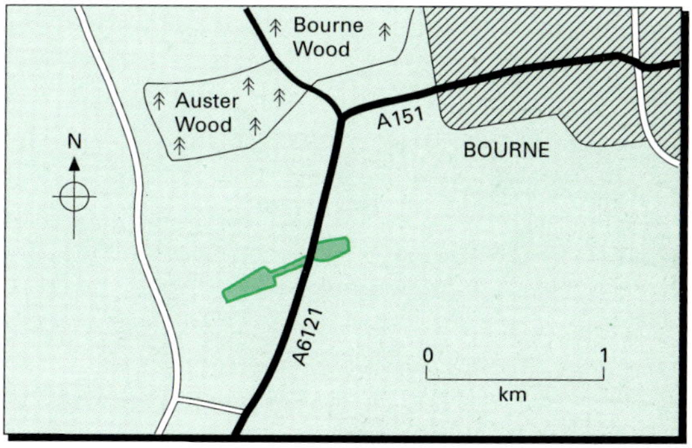

Tortoiseshell Wood and Meadows

North Witham
OS: 130 GR: SK 963197 21.2 hectares (52.6 acres)
Freehold 1972
Map ref: 57

Location and Access

The reserve lies about 2.5 km (1.5 miles) east of the A1 road some 15 km (9.4 miles) south of Grantham. From the A1 take the Castle Bytham road, approximately 5 km (3.1 miles) south of Colsterworth. The reserve entrance is about 2 km (1.25 miles) from the A1 on the north side of the road through a field gate. Park on the roadside and follow the track on foot to enter the meadows reserve known as Porter's Lodge Meadows. To access the wood, keep going straight along the track. Visitors are requested to keep to the waymarked route in the wood. Wheelchair access to the meadows is possible, but the woodland paths are not suitable.

Description and Management

The meadows, which have been traditionally managed, possess a rich and varied flora, characteristic of old calcareous clay meadows, including a wide range of grasses, abundant bird's-foot-trefoil, rough hawkbit, bulbous buttercup, cowslip, oxeye daisy, common knapweed, meadow vetchling and, in the damper furrows, water avens and cuckooflower. There are also rarer meadow plants: adder's-tongue, pepper-saxifrage, betony, great burnet, saw-wort and common cow-wheat. A small pond adjoining the wood supports several aquatic and marginal plants. The meadow/woodland edge habitat is one of the best remaining examples in Lincolnshire.

The entrance to the wood is through a small paddock, with much less floristic variety than the meadows but which is being improved by grazing. Tortoiseshell Wood is a fine example of Kesteven boulder clay primary woodland. The calcareous clay gives rise to a rich and varied ground flora. There is a typical three-layer coppice-with-standards structure. The predominant standard trees are oak and ash. The coppice consists of ash, hazel and field maple. Some of the oaks are over 150 years old and are exceptionally fine specimens. The wood is notable also for the large number of fine wild service trees. In spring and early summer the ground flora is dominated by carpets of bluebell and dog's mercury, with large scattered clumps of primrose, wood anemone and several species of violet. Enchanter's-nightshade, sweet woodruff, yellow archangel, ramsons, wood-sorrel, wood melick, wood millet, sanicle, herb-Paris, twayblade, butterfly-orchid and early-purple orchid are all to be found. Among birds are green and great spotted woodpeckers, nightingale and nuthatch — a rare species in Lincolnshire confined largely to the south-west corner of the county. Fallow deer are frequently seen in the wood.

Traditional coppice management on a 15 to 20-year rotation is being perpetuated in the main central part of the wood, while the eastern side and the perimeter are managed as high forest. There are special management difficulties associated with the restoration of coppice working. Re-growth from neglected coppice stools is poor, and deer grazing, together with invasion by brambles, exacerbates the problem.

Tortoiseshell Wood and Meadows

GR at E SK 963197

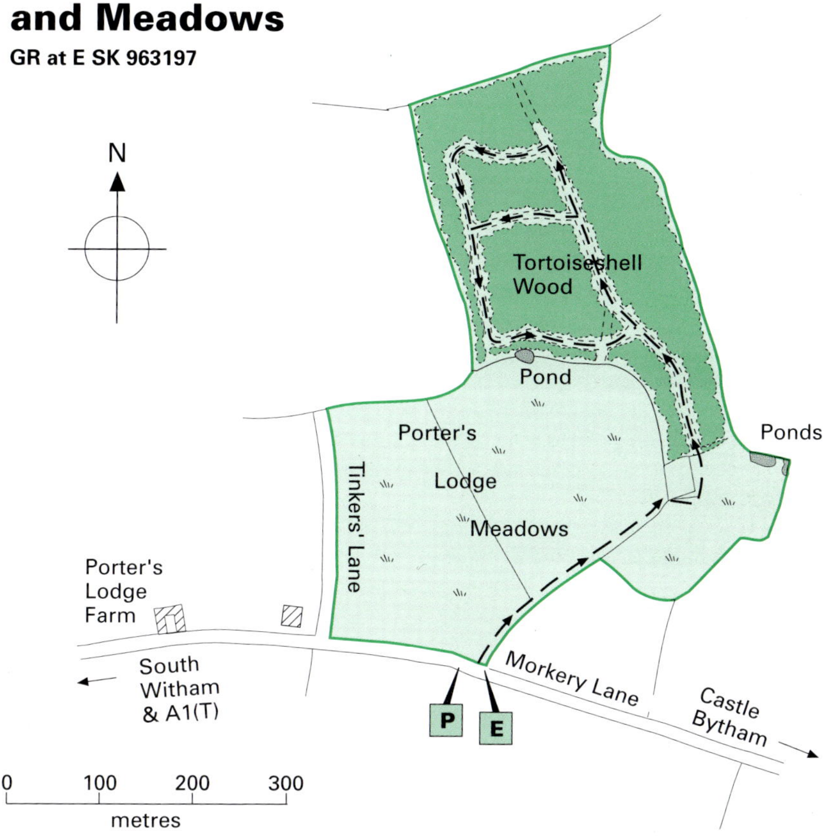

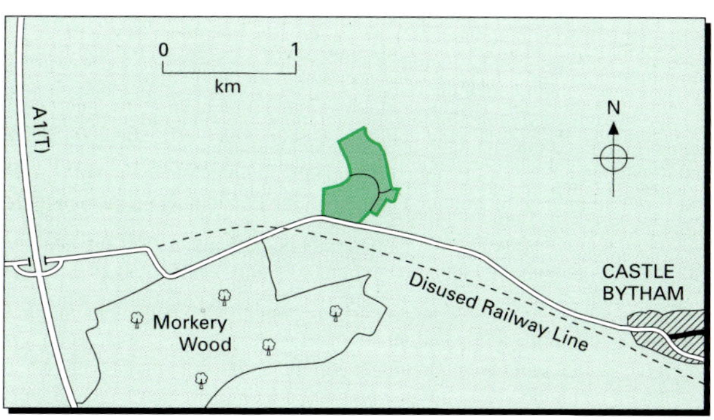

Westhorpe Pit

Gosberton
OS: 131   GR: TF 220311   0.4 hectares (0.9 acres)
Management agreement 1970
Map ref: 58

Location and Access

Situated in the hamlet of Westhorpe some 2 km (1.25 miles) west of Gosberton, the reserve lies between the railway and the road.

Description and Management

Westhorpe Pit is a small silt pit with a reedbed, where moorhens and mallard nest. Adjacent blackthorn scrub and willow provide shelter and nesting sites for tits, wren and thrushes. A small marshy area supports great willowherb, reed sweet-grass and yellow iris. The calcareous grassland on the bridge embankment supports traveller's-joy, which is a scarce plant in the Fens.

The reserve is used for educational purposes and managed by the Trust's local Wildlife Watch group.

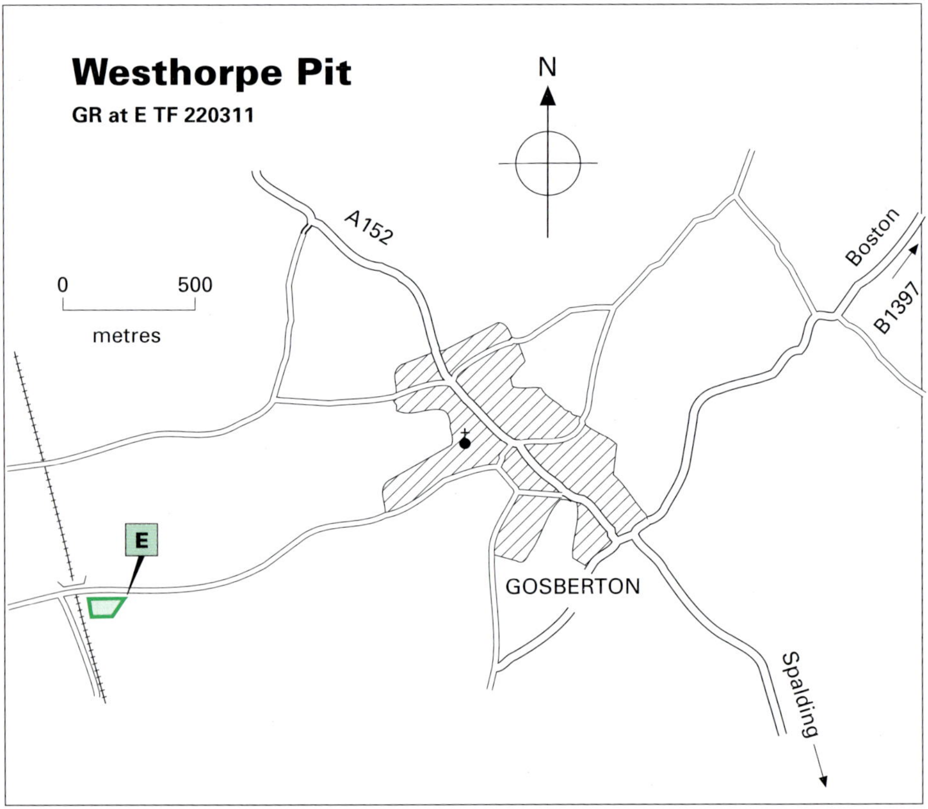

Whisby Nature Park

Doddington and Whisby, and Thorpe-on-the-Hill
OS: 121 GR: SK 914661
Nature Reserve: 20.7 hectares (51.2 acres) Nature Park: 83 hectares (205 acres)
Part freehold, part management agreement, part leasehold 1985
Map ref: 69

The Whisby Nature Park was opened in 1989 and consists of the access track from the visitor centre leased by North Kesteven District Council, land between the Pike Drain and the railway owned by the Trust, and Thorpe Lake and land north of the railway leased by the Trust. The Nature Park is managed in partnership with Lincolnshire County Council and North Kesteven District Council. Whisby Pits Nature Reserve, established in 1985, is managed by the Trust by agreement with the owner.

Location and Access

The Nature Park and the Whisby Pits Nature Reserve lie to the west of the Lincoln relief road. From the roundabout on the A46 at the southern end of the relief road, take the first turning to the left (or, if travelling southwards on the relief road, turn right after passing over the railway) into Moor Lane. The main entrance to the Nature Park, public car park and information centre will be seen on the right after approximately 0.4 km (0.25 miles). The Nature Park is indicated by brown tourist signs. There are public toilets adjacent to the information centre.

Description and Management

The Whisby Pits are the result of the excavation of sand and gravel, which were deposited in the valley of the River Witham by the retreating ice cap during the last glaciation. Some of the other pits have been developed for fishing, sailing and water skiing, but the major part of the complex of pits and former processing areas lying to the west of the Lincoln relief road have been set aside for wildlife and quiet public enjoyment.

There are two full-time wardens for the Nature Park. Management includes extensive work to maintain wetland, grassland and scrub habitats, and path mowing, monitoring and visitor services. A circular path has been created at the northern end of the Nature Park and extensive improvements to habitats are being undertaken here, including the creation of a sand martin cliff.

The Nature Park

Whisby Nature Park has been designed to allow as many visitors as possible to see and enjoy wildlife associated with a wide variety of habitats. A small information centre with interpretive displays is located at the entrance on Moor Lane. Many attractive walks of varying lengths are described in the displays and are waymarked. The area is especially suitable for school groups for whom special information and tuition is available by arrangement. From the car park follow the perimeter track, which leads to a foot-bridge across the Pike Drain. From the bank good views over the lakes lying between the Pike Drain and the railway may be had. To the right of the bridge are three birdwatching hides. In winter many wildfowl are to be seen, including wigeon, teal, tufted duck, pochard and goldeneye. In summer great crested grebes, mallard and kingfisher nest, and common terns nest on rafts provided for them. Many wading birds stop off on migration. In the copses and scrub many other nesting birds may be seen, including reed and sedge warblers, whitethroat and lesser whitethroat, nightingale, green woodpecker and willow tit. There is a foot-bridge over the railway that leads to another part of the Nature Park, leased by the Trust, consisting of small flooded pits, marsh, willow carr and scrub.

Whisby Pits Nature Reserve

The Nature Reserve is subject to a management agreement with the owners, Robert Teal Ltd. It is not open to the public except by special permit. To reach the reserve, approach from the south, taking the route from the information centre through the Nature Park. From the railway footbridge take the path northwards. From here the route is waymarked to the shore of the lake where a bird hide is planned for the future. The waymarked path then follows a circular route back to the entrance of the Nature Park. Visitors are requested not to enter the working area of the site.

The lakes and islands, with fringing scrub of willow, sallow and hawthorn, beds of reedmace and open grassy glades, make the Whisby Pits a place full of interest for wildlife. Nearly 450 species of plants and over 180 species of birds have been recorded. It is also rich in invertebrates, particularly dragonflies, of which 20 species have been recorded, and moths, of which 250 species have been recorded.

View from Den's Den, Whisby Pits

Whisby Pits and Nature Park
GR at E SK 914661
WHISBY
Sand Hills
Whisby Moor Farm
Nature Reserve
H
Nature Park
Railway
Nature Park
Pike Drain
Thorpe Lake
Pike Drain
Lincoln Relief Road A46
N
Ski World
P
E
Moor Lane
0
250
500
metres
N
WHISBY
A46(T)
Railway
Lincoln
A1434
THORPE-ON-THE-HILL
A46(T)
0
1
km

Willoughby Branch Line

Willoughby and Farlesthorpe
OS: 122 GR: TF 474736 to 467719 4.9 hectares (12.2 acres)
Leasehold 1982
Map ref: 63

Location and Access

The main access to the reserve is on the Farlesthorpe–Willoughby road. A small car park is provided near this entrance. From the B1196 road, a few metres east of the old Willoughby station crossing, a green lane access may be used by courtesy of the owner. A public path runs the length of the reserve. The reserve is leased from Lincolnshire County Council and is a statutory Local Nature Reserve. For a location plan with adjoining reserves see page 51.

Description and Management

The reserve is part of the disused branch railway from Willoughby to Mablethorpe. Since the line was taken up in 1971, the track has developed into a fine wildlife area with ashwood, hawthorn scrub and grassland supporting a varied flora, including abundant bird's-foot trefoil and restharrow, hemp-agrimony, spotted-orchid, twayblade, lady's bedstraw, yellow-wort and great burnet. Butterflies include common blue and several species of skippers and browns. Whitethroat, lesser whitethroat, blackcap, sedge warbler, redpoll and other finch species, and occasionally nightingale, nest in the reserve. Hips and haws attract fieldfares and redwings in winter, and barn owls frequently use the track for hunting.

Management consists of twice-yearly mowing of the path and the taking of a crop of hay from the wider tract of grassland at the Farlesthorpe end of the reserve. Oak, field maple, birch and alder have been planted in suitable sites along the track. Hawthorn bushes are coppiced occasionally to prevent them overgrowing the path.

Barn owl

Willoughby Meadow

Willoughby-with-Sloothby
OS: 122 GR: TF 472713 0.5 hectares (1.3 acres)
Freehold 1976
Map ref: 60

SSSI

Location and Access

The reserve is situated 0.5 km (0.3 miles) south of Willoughby: from the B1196 road turn off for Sloothby at the crossroads down Mill Lane, and the reserve is less than 100 m on the right. As this small meadow is vulnerable to trampling, especially before hay time, visitors are requested to avoid damaging the crop. For a location plan with adjoining reserves see page 51.

Description and Management

This small meadow on the Middle Marsh Boulder Clay is one of the finest remaining examples of its type. There is a rich assemblage of meadow plants, including cowslip, great burnet, pepper-saxifrage, adder's-tongue, devil's-bit scabious, saw-wort, betony and green-winged orchid. Bugle and lady's mantle, more usually associated with woodland rides, also occur. No less than 149 species of flowering plants have been recorded. There are good boundary hedges and old hedgerow trees attractive to whitethroat and lesser whitethroat.

Management usually entails hay-cutting in mid- to late July, followed by grazing for a few weeks in autumn.

Devil's-bit scabious (left) and pepper-saxifrage (right)

Witham Way

Bardney to Boston
OS: 121,122 and 131
GR: TF 117585 (Bardney) to TF 317455 (Boston) (see notes below)
*c.*100 hectares (*c.*250 acres)
Management agreement 1990
Map ref: 87

Location and Access

The reserve consists of disused sections of the former Boston–Lincoln railway on the bank of the River Witham, between Bardney and Boston. It is hoped that it may eventually become possible to link existing sections of the reserve into a wildlife corridor, but at present access to the Witham Bank is rather restricted. In some places (e.g. Dogdyke to Langrick) there is no public right of way, nor any access for visitors to Trust-managed land, except by the consent of tenants or owners of adjacent land. For this reason advice should be sought from the Trust before attempting to visit the reserve between Dogdyke and Langrick, and, in particular, the Swinton's Farm section of the Witham Bank south of Dogdyke. Specific access and location of each of the sections is detailed below.

Description and Management

The nature reserve is in three sections:

Bardney (TF 117585) to Kirkstead (TF 188602)
(11.5 km: 7.1 miles)

There is an access at the former Woodhall junction near Kirkstead Bridge at grid reference TF 621176. Parking space is available at the roadside. The walk northwards past the old Southrey railway station to the Bardney sugar beet factory is particularly rich, with well developed scrub and grassland habitats. There are also good access points at Southrey station (where there is a car park: TF 139664) and at Stixwould station (parking is at the former station yard at TF 159649).

Dogdyke (TF 210555) to Langrick (TF 262480)
(9 km: 5.3 miles)

Only the central part of this section is in Trust management, and there is no public right of way or automatic right of access for visitors from either end of the Witham Bank. Some sections are privately tenanted.

Langrick Bridge (TF 264476) to Boston (TF 322447)
(6.5 km: 4 miles)

There are access points at Langrick Bridge, Anton's Gowt (TF 300475) and Boston.

There are 21 species of butterflies recorded on the track, river banks and scrub of the old railway, including wall, hedge brown, gatekeeper and common blue. The flora is varied, with such species as scarlet pimpernel, common spotted-orchid, twayblade, whitlowgrass, stonecrop and tufted vetch. In marshy places beside the river, and in the ponds, false fox-sedge, hairy sedge, water-dropwort, reed and great reedmace grow. The old hedges and scrub, containing hawthorn and blackthorn with hazel, crab apple, oak, ash, sweet briar, dog rose and alder, attract many warblers and finches. Toads, frogs and newts occur, and there have been recent sightings of otters in and around the River Witham. The river, which runs beside the route for most of its way, attracts ducks and grebes. Herons are also frequent visitors.

Management consists chiefly of scrub control, path maintenance and the mowing of open grassland habitats.

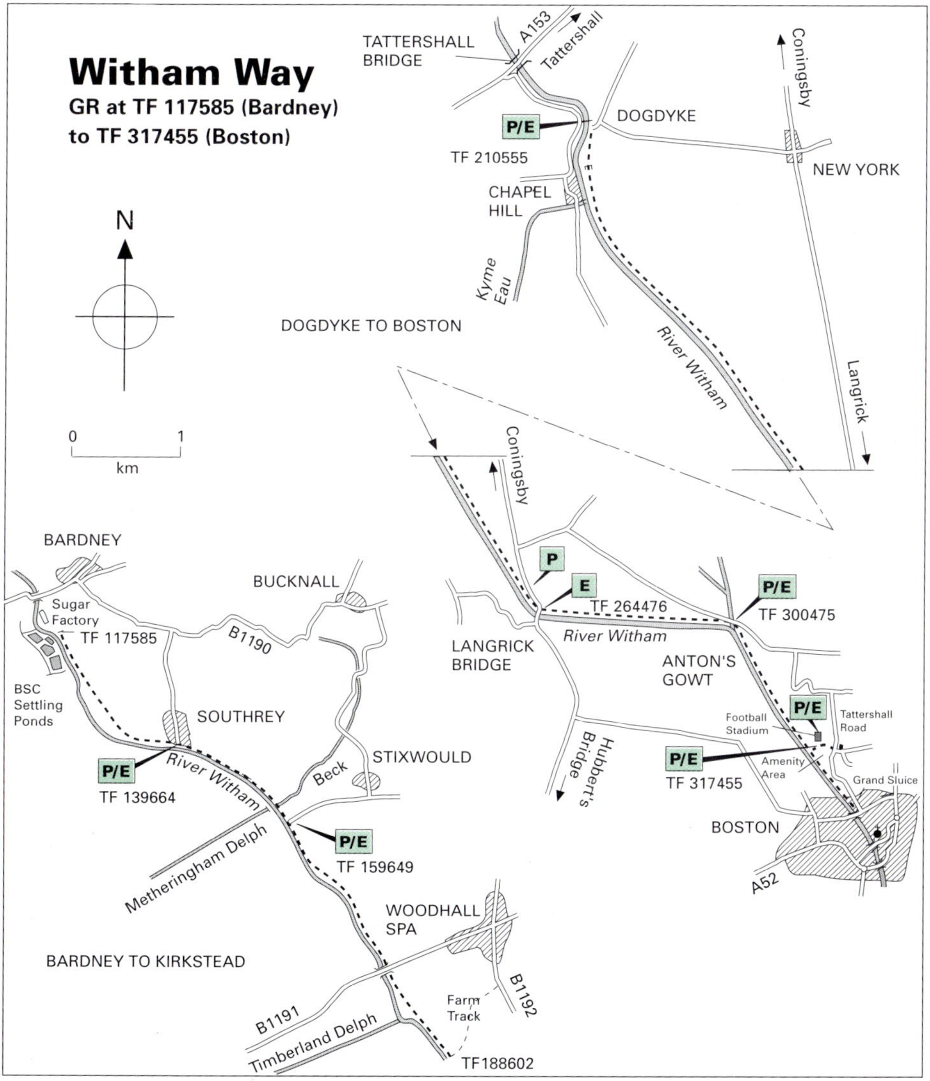

Wolla Bank Pit

Chapel St Leonards
OS: 122   GR: TF 555747   3.9 hectares (9.6 acres)
Leasehold 1959
Map ref: 46

S SSSI

Location and Access

Close to the sea wall about 1.6 km (1 mile) south of Anderby Creek, the reserve lies on the west side of the road opposite the entrance to the Wolla Bank car park. Visitors are requested not to walk round the pit to avoid disturbing wildfowl: good views can be had by walking down the short path from the road. For a reserve plan see map opposite, and for a location plan with adjoining reserves see page 49.

Description and Management

The flooded pits are a result of excavations for clay, which was needed for the repair of the sea wall following the floods of 1953. There are extensive beds of reed and sea club-rush, with great reedmace, fennel pondweed, wild celery, sea arrowgrass and water-crowfoot. There are also colonies of common spotted-orchid. Many duck visit the reserve in winter, including tufted duck, goldeneye and pochard. Bewick's and whooper swans, red-necked, slavonian and great crested grebes, purple heron and penduline tit have all occurred. Snipe are frequent visitors in winter, when bittern and bearded tit are occasional visitors. In the breeding season, reed and sedge warblers (and occasionally grasshopper warblers), reed bunting, and great crested and little grebe all nest. Many rare migrants have also been seen. The entrance track made of chalk has a range of chalkland flora and many butterflies, such as common blue, small copper and wall. Dragonflies, such as migrant hawker, also occur. The Wolla Bank Reedbed lies on the opposite side of the road.

Snipe

Wolla Bank Reedbed

Anderby
OS: 122 GR: TF 556749 2.9 hectares (7.1 acres)
Leasehold 1965
Map ref: 45

S SSSI

Location and Access

The reserve lies between the coast road and the sea wall about 1.6 km (1 mile) south of Anderby Creek. There is no access to the reserve, which can be viewed from the Wolla Bank car park and access road. For a location plan with adjoining reserves see page 49.

Description and Management

Having been partially scraped for clay for the repair of sea defences following the east-coast floods of 1953, the area is now an extensive reedbed. Reed warbler, sedge warbler, reed bunting and whitethroat all nest. Bearded tits are occasional visitors in winter. Marsh harriers occur in spring and short-eared owls in winter. There are small pools and grassy areas with sea club-rush and sea couch. Small stands of sea-buckthorn occur here and in the Wolla Bank Pit, which lies on the opposite side of the road. Sea buckthorn attracts wintering thrushes when in fruit.

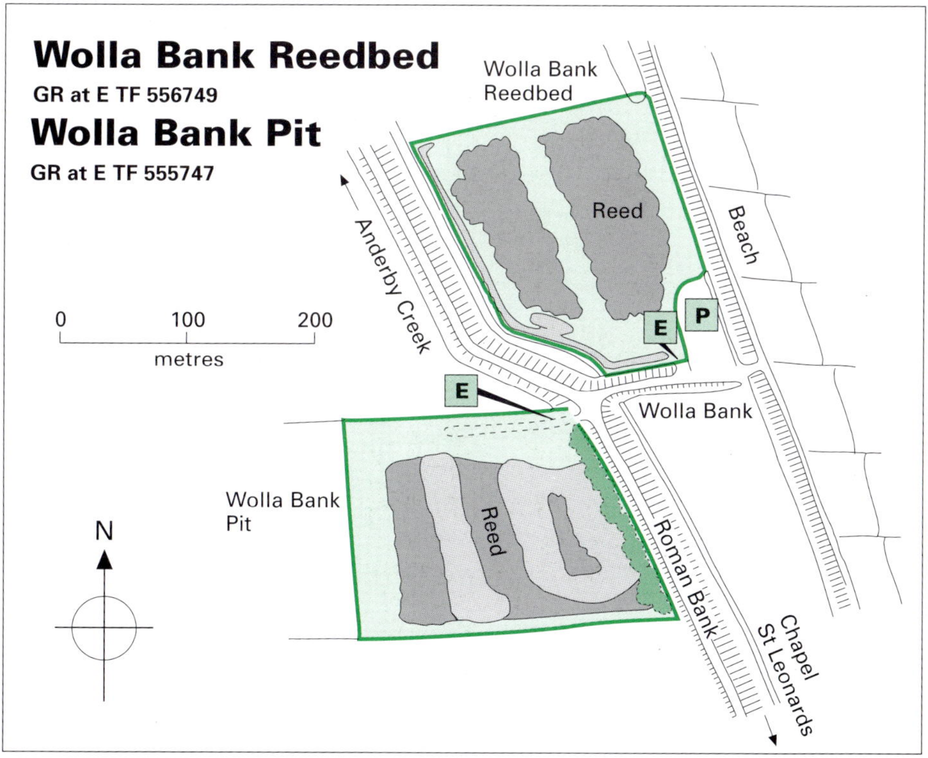

Woolsthorpe Line

Colsterworth
OS: 130 GR: SK 929249 to 902242 7.7 hectares (19 acres)
Management agreement 1990
Map ref: 91

Location and Access

This 3 km length of disused railway line lies to the west of Colsterworth, near Grantham, passing close to the birthplace of Sir Isaac Newton at Woolsthorpe-by-Colsterworth. There are several access points from the nearby minor road and, at the eastern end, access may be had from the embankment on the A1 slip-road. The section west of the Stainby–Skillington crossroads is suitable for wheelchair access.

Description and Management

At the eastern end there is well developed scrub and woodland edge habitats, and the route adjoins attractive meadows that run down to the tree-lined River Witham. There is a good range of breeding birds, including willow warbler, blackcap, linnet and stock dove, while in winter the scrub attracts feeding flocks of finches, fieldfare and redwing. To the west there is less scrub, but hedgerows, rough grassland and tall herb provide an attractive walk. A variety of butterfly species can be seen in the summer along the section west of the Stainby–Skillington crossroads. A protected roadside verge (PRV 048) runs parallel to the reserve, eastwards from the crossroads. Bee orchids occur on the southern verge.

Management consists of scrub control and regular mowing of the grassland areas.

Stock dove

Woolsthorpe Line

GR at SK 929249–902242

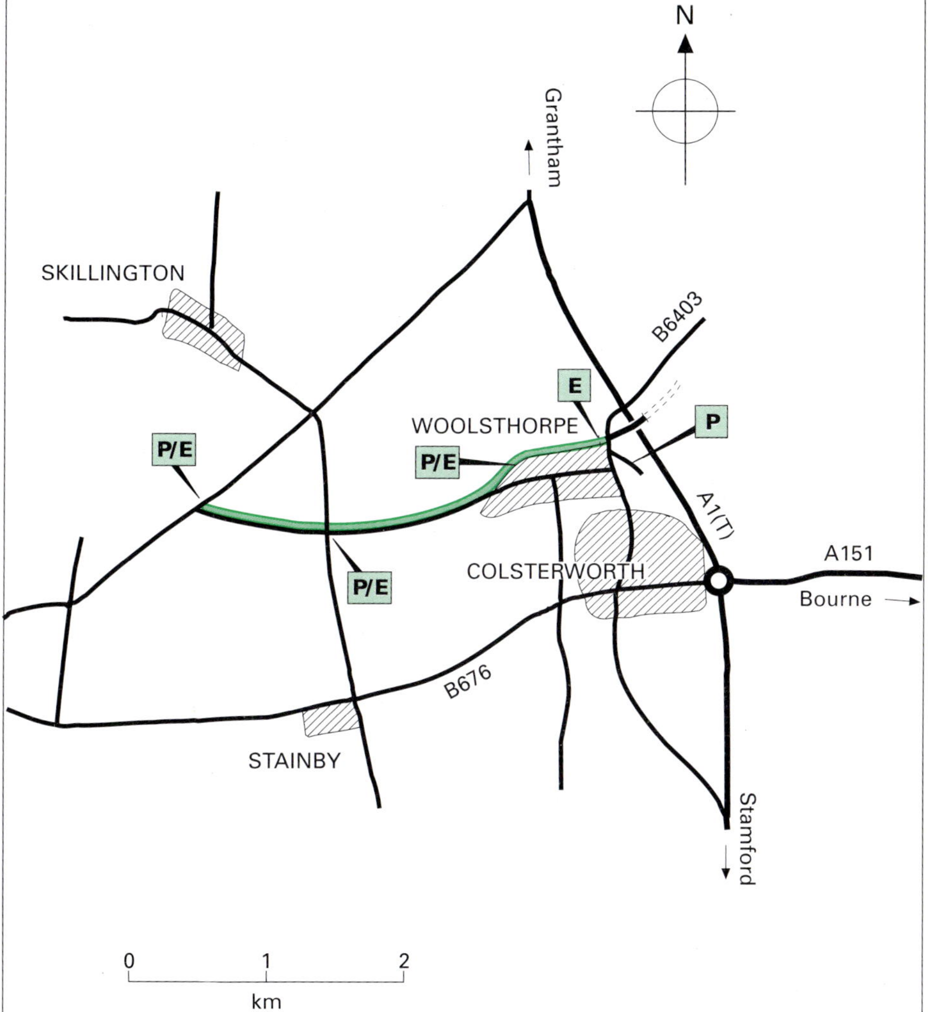

Protected Roadside Verges

One of the most pleasing aspects of the Lincolnshire countryside is the wide verges along many of the roadsides. They are havens for a variety of invertebrates and small mammals, and in summer they are coloured with many kinds of wildflowers. The verges of the county together form the largest remaining tract of semi-natural vegetation. Intensive agricultural practices, along with increased industrial and housing development, have threatened the survival of wildlife and wild places. The verges are therefore of great importance in providing a reservoir of plants and animals, some of which have become scarce or localised owing to the loss of suitable habitat elsewhere.

In Lincolnshire the richest verges are found on the alkaline soils of the chalk and limestone where they contain such rare plants as yellow-wort, autumn gentian, dyer's greenweed, clustered bellflower, pyramidal and man orchids. Some verges on clay soils have a varied meadow flora, including such species as cowslip, spotted-orchid and adder's-tongue. There are interesting verges also on the more acid sands and gravels in the centre and north-west of the county.

Because of the special role that verges play in wildlife conservation, and because of the need to have regard to factors that might change their character (and therefore their biological interest and attractiveness), a scheme for their protection and management was set up in 1960. The scheme is run in cooperation with Lincolnshire County Council whose Highways Division provides financial and advisory support. Management is carried out by the Trust in accordance with agreed plans, which take account of wildlife needs and road safety.

There are 55 Protected Roadside Verges, covering either both sides of the road or only one side, which total a distance of 74 kilometres (46 miles). For each verge, the Trust appoints a voluntary 'Wayside Warden' to help look after the biological interest in liaison with the Divisional Surveyors and landowners.

When visiting the verges please take care to park safely and not to park on the protected verge. Take special care to look out for traffic.

Ref. no.	**Grid reference**	**Name/Parish**	**Side of road**
001	TF 193752–197758	Baumber	Both
002	TF 215924–220908	Binbrook	Both
003	TF 322761–328762	Bluestone Heath	South
004	TF 386757–390756	Calceby	Both
005	TF 449713–451709	Claxby	Both
006	TF 452701–458708	Dawber Lane	Both
007	TF 225629–227636	Wellsyke Lane (see page 139)	Both
008	TF 401766–399756	South Thoresby	Both
009	TF 128879–136875	Linwood (see page 124)	South
010	TA 139000–TF 143998	Nettleton	Both
011	TF 009147–006156	Mill Farm, Holywell	East

Ref. no.	Grid reference	Name/Parish	Side of road
012	TF 262805–265809	Red Hill	Both
013	TF 432752–433749 & TF 419762–423759	Rigsby	Both
014	TF 291784–295785	Rowgate Hill	Both
015	SK 866982–874985	Scotton (see page 164)	Both
016	TF 328762–329755	Tetford Hill	Both
017	TF 060963–063963	Thornton-le-Moor	Both
018	SK 846789–849786	Torksey	North
019	SK 951349–953353	Blue Harbour	East
020	SK 989487–997488	Byard's Leap Gorse	Both
021	SK 989486–990493	Byard's Leap Plantation	Both
022	TF 028437–036443	California Plantation	South
023	SK 996172–997169	Castle Bytham	Both
024	SK 982433–974424 & 978421 & 979426	Copper Hill (see page 65)	Both
025	TF 023486–023478	Hanthorpe	Both
026	SK 990495–991513	High Dyke	Both
027	SK 946333–948341	Old Somerby	East
028	TF 006160–009161	Holywell to Careby	Both
029	TF 010144–027137	Little Warren	Both
030	SK 960214–970222	Lobthorpe	Both
031	TF 033608–050613	Metheringham Heath	Both
032	TF 419776–423775	Swinn Wood	Both
033	TF 455686–457687	Mill Lane	Both
034	TF 009431–012430	Wilsford Bank	North
035	TF 076994–078995	Moortown	South
036	TF 372805–375810	Burwell Wood	Both
037	TA 164111–167113	Roxton Wood	Both
038	SK 954193–955194	Morkery Wood	South
039	TF 337763–42765	Worlaby	South
040	SK 931251–33253	Colsterworth Bank	Both
041	TF 025346–32348	Sapperton	Both
042	SK 894874–03868	Upton Grange	Both
043	TF 004599–11602	Boothby Graffoe Heath	South
044	TF 285635–291639	Moorby	Both
045	TF 338920–40916	Brackenborough	East
046	TF 196672–98669	Furze Hill	Both
047	SK 988157–TF 003158	Holywell Wood	Both
048	SK 908242–917243	Woolsthorpe	South
049	SK 964346–973347	Ropsley	Both
050	SK 857652–862658	Eagle	East
051	SK 931191–945193	South Witham	Both
052	TF 281735–289732	Fulletby/West Ashby	Both
053	TF 146753–158767	Hoop Lane	Both
054	SK 996233–TF 004233	Eager Farm	North
055	TF 308659–310659	Hammeringham Hill	Both

Index of Parishes

PRV = Protected Roadside Verge (see page 208)

Parish	Reserve/PRV	Map no.	Page no.
Claxby-by-Alford	Hoplands Wood	25	100
	Skendleby Psalter Banks	48	168
	PRV 005		
Colsterworth	Woolsthorpe Line	91	206
	PRV 040, PRV 048		
Covenham St Bartholomew	Covenham Reservoir	11	52
Cranwell	PRV 020, PRV 026		
Croft	Gibraltar Point NNR	22	84
Crowland	Crowland Ponds	98	54
Crowle	Crowle Moor	12	56
Digby	Digby Corner	94	60
Doddington and Whisby	Whisby Nature Park	69	197
Donington	The Yews	56	189
Dunholme	Pickering's Meadow.	70	146
East Keal	Keal Carr	65	106
East Lound	Sedge Hole Close	26	166
Epworth	Epworth Turbary	17	67
Ewerby and Evedon	Ewerby Pond	18	70
Farlesthorpe	Willoughby Branch Line	63	200
Fishtoft	Frampton Marsh	19	76
	Hobhole Bank	89	98
Friskney	Friskney Decoy Wood	20	80
Glentworth	PRV 042		
Grimsby	Freshney Bog	73	78
Gosberton	Boston Road Brick Pits	7	40
	Westhorpe Pit	58	196
Goulceby	Red Hill	37	150
	PRV 012		
Goxhill	Dawson City Clay Pits	76	59
Hagworthingham	Furze Hill	21	82
Hatton	Hatton Meadows	83	92
Haxey	Axholme Line — Haxey	1	30
	Haxey Turbary	23	94
	Langholme Wood	29	118
	Rush Furlong	40	158
Heapham	PRV 004		
Holywell	PRV 028, PRV 029		
Horbling	Horbling Line	72	103
Horncastle	Banovallum House	100	32
Horsington	PRV 046		
Huttoft	Huttoft Bank Pit	44	104
Kirkby-la-Thorpe	Kirkby-la-Thorpe Pit	84	117
Kirkby-on-Bain	Kirkby Moor	28	114
	Moor Farm	35	137
	PRV 007		
Langton-by-Wragby	Little Scrubbs Meadow	31	127
	Goslings Corner Wood	79	88
Lincoln	Boultham Mere	81	42
Linwood	Linwood Warren	30	124
	PRV 009		

Parish	Reserve/PRV	Map no.	Page no.
Tattershall Thorpe	Kirkby Gravel Pit	92	112
	Kirkby Moor	28	114
Tetford	PRV 003, PRV 016		
	PRV 039		
Tetney	Tetney Blow Wells	54	184
Theddlethorpe	Saltfleetby–Theddlethorpe Dunes NNR	41	159
Thorpe-on-the-Hill	Whisby Nature Park	69	197
Thurlby	Dole Wood	14	61
	The Chasm & Northorpe Slipe	80	187
	Thurlby Fen Slipe	82	190
Torksey	PRV 018		
Toynton All Saints	Keal Carr	65	106
	Lilley's Carr	95	122
Wainfleet St Mary	Gibraltar Point NNR	22	84
Well	PRV 013		
Welton-le-Marsh	Candlesby Hill Quarry	9	46
	Dawber Lane Quarry	13	58
	PRV 033, PRV 006		
Willoughby	Willoughby Branch Line	63	200
Willoughby-with-Sloothby	Willoughby Meadow	60	201
Wilsford	Duke's Covert & Copper Hill	71	65
	PRV 034		
Winceby	Snipe Dales	49	169
Witham-on-the-Hill	Stanton's Pit	103	178
Wyberton	Frampton Marsh	19	76

Index of Reserves

A Living Legacy

The nature reserves described in this handbook are among the most attractive places in Lincolnshire and are home to a great variety of wild plants and animals, for many of them a last refuge.

The Lincolnshire Trust for Nature Conservation continues to strive to protect wildlife and wild places throughout the county against destruction, pollution and neglect. By making a bequest to the Trust in your will, you can help to ensure that Lincolnshire's natural heritage is safeguarded for future generations. You can leave to the Trust a specified cash sum or the residue of your estate or a specified share of it. Land or other property or belongings may also be bequeathed, but it is advisable to discuss such intentions with the Trust beforehand. Whatever your decision, we strongly recommend that you consult your solicitor when making a will or changing an existing one. The Trust will be pleased to discuss the provision of a legacy with you or with your solicitor, and any contacts will be treated in the strictest confidence. For further information on leaving a legacy in your will, a leaflet is available from the Trust's headquarters: The Lincolnshire Trust for Nature Conservation, Banovallum House, Manor House Street, Horncastle, Lincolnshire LN9 5HF.

75% RECYCLED

The text paper in this handbook contains 50% Post Consumer White Waste, 25% Packaging Waste and 25% Virgin Pulp. All these items are bleached using a chlorine-free oxygen based process. If the waste items were not recycled they would have been destined for a land fill site.

The cover paper uses Virgin Pulp made from saw mill residues, forest thinnings and trees cut from sustainable forests in Finland. The mill waste is recycled and makes up about 20% of the total. Once again it is elemental chlorine free with an AOX level of < 0.5kg per tonne of pulp. Strict local laws mean that the water used in the mill is very often returned to the Eco-system in a better state than when it was removed.